OXFORD ENGINEERING SCIENCES SERIES

THE OXFORD ENGINEERING SCIENCE SERIES

1 D. R. RHODES: *Synthesis of planar antenna sources*
2 L. C. WOODS: *Thermodynamics of fluid systems*
3 R. N. FRANKLIN: *Plasma phenomena in gas discharges*
4 G. A. BIRD: *Molecular gas dynamics*
5 H.-G. UNGER: *Planar optical waveguides and fibres*
6 J. HEYMAN: *Equilibrium of shell structures*
7 K. H. HUNT: *Kinematic geometry of mechanisms*
8 D. S. JONES: *Methods in electromagnetic wave propagation (Two volumes)*
9 W. D. MARSH: *Economics of electric utility power generation*
10 P. HAGEDORN: *Non-linear oscillations (second edition)*
11 R. HILL: *Mathematical theory of plasticity*
12 D. J. DAWE: *Matrix and finite element displacement analysis of structures*
13 N. W. MURRAY: *Introduction to the theory of thin-walled structures*
14 R. I. TANNER: *Engineering rheology*
15 M. F. KANNINEN and C. H. POPELAR: *Advanced fracture mechanics*
16 R. H. T. BATES and M. J. McDONNELL: *Image restoration and reconstruction*
18 K. HUSEYIN: *Multiple-parameter stability theory and its applications*
19 R. N. BRACEWELL: *The Hartley transform*
20 J. WESSON: *Tokamaks*
21 P. B. WHALLEY: *Boiling, condensation, and gas-liquid flow*
22 C. SAMSON, M. LeBORGNE, and B. ESPIAU: *Robot control: the task function approach*
23 H. J. RAMM: *Fluid dynamics for the study of transonic flow*
24 R. R. A. SYMS: *Practical volume holography*
25 W. D. McCOMB: *The physics of fluid turbulence*
26 Z. P. BAZANT and L. CEDOLIN: *Stability of structures: principles of elastic, inelastic, and damage theories*
27 J. D. THORNTON: *Science and practice of liquid–liquid extraction (Two volumes)*
28 J. VAN BLADEL: *Singular electromagnetic fields and sources*
29 M. O. TOKHI and R. R. LEITCH: *Active noise control*

Active Noise Control

M. O. TOKHI

*The University of Sheffield,
Department of Automatic Control and Systems Engineering,
Sheffield, UK*

R. R. LEITCH

*Heriot–Watt University,
Department of Electrical and Electronic Engineering, Edinburgh, UK*

CLARENDON PRESS • OXFORD

1992

Oxford University Press, Walton Street, Oxford OX2 6DP

Oxford New York Toronto
Delhi Bombay Calcutta Madras Karachi
Petaling Jaya Singapore Hong Kong Tokyo
Nairobi Dar es Salaam Cape Town
Melbourne Auckland

and associated companies in
Berlin Ibadan

Oxford is a trademark of Oxford University Press

Published in the United States
by Oxford University Press, New York

A catalogue record for this book is available from the British Library

Library of Congress Cataloging in Publication Data
Tokhi, M. O.
Active noise control / M. O. Tokhi, R. R. Leitch.
(Oxford engineering science series ; 29)
Includes bibliographical references and index.
1. Noise control. 2. Acoustic impedance. I. Leitch, R. R.
II. Title. III. Series.
TD892.T65 1992 620.2'3—dc20 91-24762
ISBN 0-19-856243-8

Typeset by APS, Salisbury, Wiltshire
Printed in Great Britain by St Edmundsbury Press Ltd
Bury St Edmunds, Suffolk

TO

Farida, Ihsanulla, and Meerwais Tokhi

and

Marjory, Stuart, and Karen Leitch

PREFACE

The prospect of the active control of noise has intrigued, and frustrated, scientists and engineers in many countries for over five decades. The basic idea is relatively simple: to generate artificially a secondary acoustic field that destructively interferes with the field produced by the unwanted sound. However, the limitations of the available electronic technology have restricted practical realization to a few simple applications. With the development of flexible and cheap digital signal processing devices, capable of processing signals within the audio range, there has been resurgence in the interest in the theory and practice of developing Active Noise Control (ANC) systems. Further, the flexibility of these devices allows the development to concentrate on the design procedure and corresponding control algorithms, rather than the acoustic properties of the arrangement. This book provides a principled introduction to the design of ANC systems from the perspective of control engineering. The approach allows the designer to predict the performance of the system in advance and reduces the empirical aspect of many previous approaches. The use of control engineering methods allows the progressive development of various control techniques, from fixed digital control to self-tuning adaptive control, to be realized within a single coherent framework.

The content of the book reports research and development work undertaken by the authors, within the Department of Electrical and Electronic Engineering, Heriot-Watt University. During this time Dr Tokhi was supported by the British Council Research Fellowship, which is gratefully acknowledged. Both authors are indebted to the support and encouragement of their families; their patience and understanding during the construction of this research and its subsequent reporting have been crucial to its successful completion.

The authors would like to acknowledge the enthusiastic encouragement and support of Professor Michael Brady (University of Oxford) and Professor Derek A. Linkens (University of Sheffield). Furthermore, many thanks to the staff of the Oxford University Press for their encouragement, advice, and patience.

Sheffield M.O.T
Edinburgh R.R.L.
September 1991

CONTENTS

1
INTRODUCTION

Acoustic noise has long been recognized as a source of environmental pollution, that can have adverse effects on human life in numerous ways. Attempts have been made to solve this problem by improving the design of sources of noise, so that the level of sound they emit is reduced, and/or by developing methods, external to the source, for sound attenuation. Many countries have developed legislation and standards to limit the allowable sound level, especially in the workplace.

There are a number of factors that lead to 'sound' (audible or non-audible acoustic waves) being perceived as 'noise' by humans, for example: the sound level may be too high: the sound may contain frequencies that are disturbing; the sound structure may not correspond to the person's conception; or the sound may bear an unpleasant meaning such that the person cannot or will not be receptive to it. Investigations have shown that in any of these situations the noise may affect a person psychologically and/or physiologically. Noise has been found to cause permanent damage to the inner ear with resulting permanent hearing loss that can range from slight impairment to nearly total deafness. It can result in temporary hearing loss and, by repeated exposure, can lead to chronic hearing losses with associated raised blood pressure and other physiological disorders. Noise can be a source of interference with speech, communication, and the perception of other auditory signals. It can also interfere with the performance of complicated tasks and can especially disturb performance when speech communication in response to auditory signals is demanded. Noise can adversely influence mood, disturb relaxation and sleep, and, with other acoustic considerations, can affect privacy.

One of the major considerations in specifying the effects of noise is the adaptability of the human being. Many noise-induced effects are observable only in the short term and weaken or disappear over time. The lack of observable effects following adaptation does not prevent noise from having indirect effects. People who work or perform a task under noisy conditions may show decreased abilities after the noise has ceased.

The most significant health problem caused by noise is hearing loss. Exposure to audible sound of sufficient intensity for a long period of time can produce physiological changes in the auditory system. Some of the impairments are temporary and disappear with time. A regular schedule of exposure to sounds of sufficient intensity, however, can lead to permanent loss of the actual auditory receptors. For instance, if a person is subjected to a sound level of over 85 dB (A) for several hours (eight hours, say) each day for a number of years permanent damage will occur. Similar damage to the ear may be expected at an average sound level of 80 dB (A) over a period of 24 hours. The greater the amount of sound, that is, the longer the effect and the higher the level, the greater will be the probability of noise-deafness occurring. Higher sound levels can only be supported without damage to the hearing if the exposure times are correspondingly shorter. Practical tests have shown that with impulse noise it must also be realized that the probability of damage does not depend on a physical change, but that noise impulses with peak values of 150 dB and more may be expected to cause direct mechanical damage to the inner ear.

The primary effect due to low-frequency infrasonic noise appears to be annoyance. Response to low frequency (inaudible) noise, however, has varied from sleep disturbance and general annoyance to a severe loss in concentration and fatigue. It has been found that, for decreasing frequency, a higher sound pressure level is required to provoke the same amount of annoyance. However, as the frequency decreases, the change in annoyance becomes more pronounced for a change in the noise level. Thus, in the infrasonic region, relatively small changes in sound pressure may cause large changes in annoyance. This is particularly noticeable in the cab of large diesel trucks, where the dimensions of the cab and the long hours of exposure have made infrasound a real safety problem for drivers.

Low-frequency noise in the range of 200 Hz and below has been found to cause structural vibration. For example, low-frequency sound from a large machine or vehicle near residential buildings can result in induced noise as high as 65–70 dB in linear scale and can cause glass windows of buildings to chatter, leading to the so-called secondary or induced-noise problem. This takes place by a process of coupling of the low-frequency sound into the structure via the windows and doors. In this manner, acoustic coupling can excite the window and the contents of the room into vibration. Floors and walls can also be caused to vibrate. This can be detected by the occupants of the building as vibrational responses of the building and its contents and even through direct perception as body vibrations.

The legislation on limiting community noise or occupational noise exposure is most commonly based on the presence of annoyance reactions among exposed population groups. Behind these reactions lie non-auditory health effects from moderate levels of noise (below 80 dB) which appear to be seen only after long-term exposures. The current legislation in many countries covers areas of traffic noise, occupational noise exposure, and noise as a general nuisance to public health. For each of the cases, standards and norms for maximum levels of noise and noise exposure as well as techniques of noise measurement are set. To meet these norms and standards, methods of noise attenuation suitable for the problem in hand are employed which, for each case, requires an understanding of the acoustics of the problem and, for design, an analysis from the control engineering perspective.

Acoustic wave propagation mechanisms, depending on the transmitting medium and its boundary conditions, can broadly be classified as dispersive and non-dispersive. In a non-dispersive propagation medium, the waves travel at velocities independent of frequency, whereas in a dispersive propagation medium, the wave velocities are frequency dependent leading to a dispersion of waves, containing a range of frequencies, with distance in the medium. The developments throughout this book assume a non-dispersive wave propagation mechanism.

A source of noise emitting acoustic waves into a medium produces a field which may be regarded as made up of two components, namely, a travelling-wave field, and a standing-wave field. These two components may be identified as follows. For the travelling-wave field the mean acoustic energy flow is non-zero; that is, the time average over a suitable interval, of the component pressure and component particle velocity, is non-zero. For the standing-wave field, the mean acoustic energy flow is zero; that is, the time average, over a suitable interval, of the component pressure and component particle velocity is zero. In this manner, for noise reduction purposes, the acoustic field within an enclosure excited by a sound source of finite size may be divided into three regions:

(1) the near field which lies within a few diameters of the source;
(2) the direct radiation or free field which is characterized for simple sources by a reduction in sound pressure level of 6 dB per doubling of distances from the centre of the source;
(3) the reverberant field existing at distances where multiple reflection from bounding surfaces become important. In an open space where there are no obstacles in the propagation path and therefore no reflection of the emitted noise takes place, the field will divide as (1) and (2) only.

The most commonly measured quantity in acoustics is sound pressure. The measurement is performed directly with a microphone and the signal can be analysed and displayed as a function of frequency. Another fundamental quantity of interest is sound intensity, which describes both the amount and the direction of flow of acoustic energy at a given position. Equipment for measuring sound power spectra suitable for the three different regions of an acoustic field, each having a certain characteristic, is readily available. Measurements in the near field do not require any special facilities and are thus ideal for on-site tests. The test effort, however, will be high due to the large number of measurement points and corresponding data processing required. Measurements in the reverberant field require only moderate facilities and the least amount of effort. Measurements in the free field can provide not only acoustic power spectra but also directivity information. The facility and test effort will, therefore, be relatively costly and hard to justify if directivity information is not needed.

The control of noise, either aimed at total or partial reduction of the noise, can be applied in a number of ways, for example: reduction at source; reduction at the ear; reduction throughout a volume of the propagation medium, and so on. The choice of the method employed can usually be determined by the practical situation of interest. In any case, two distinct methods are possible: passive absorption and active noise control. Passive noise control utilizes the absorption property of matter, that is, sound-absorbent materials are mounted on and/or around the source of noise. This technique has been proven to be cost-effective at higher frequencies (above 500 Hz, say). For lower frequencies, however, the bulk of the absorbent material required for a certain amount of reduction of noise, and hence the cost of necessary hardware, increases exponentially. This, further, leads to problems of maintenance and so on, making passive techniques impractical and uneconomical for low-frequency noise reduction.

Active noise control (ANC) uses the intentional superposition of acoustic waves to create a destructive interference pattern such that a reduction of the unwanted sound occurs. This is realized by a single or a number of secondary (cancelling) sources of sound, driven by electrical signals derived from the primary (unwanted) noise through detection sensors and then passed through an electronic controller, so that when the waves are superimposed upon one another they destructively interfere and reduction of the sound occurs. Most applications require a single-source single-detector configuration whereas there are situations where multiple-detector multiple-source configurations will be required. In either case, the generation of a stable destructive interference pattern is possible only for low frequencies, say less than 500 Hz, and for use within limited areas. This restriction of frequency range is complementary to the limitations of passive methods; practical solutions require a combination of passive techniques for higher frequencies (greater than 500 Hz, say) and active techniques for low frequencies (less than 500 Hz, say).

Initial research investigations in ANC date back to the early 1930s. Since then, there have been numerous theoretical and practical developments that have led to a wide range of successful applications. The area of ANC is still developing and is continually finding its way to new applications. Chapter 2 presents a retrospective review of the development of ANC systems. It gives a background to the area to date and describes a series of applications. The problems and limitations associated with the developments and applications are discussed and possible solutions are suggested.

The process of active noise cancellation utilizes the two basic principles of interference and absorption. By introducing secondary source(s) the acoustic energy in the medium is redistributed so that the noise is reduced in some areas and reinforced in other regions of the medium. Moreover, when cancellation sources are added, the existence of a primary sound field changes the acoustic impedance seen by these cancellation sources. However, the use of cancellation sources having a cardioid radiation pattern makes it possible to avoid influencing the acoustic impedance seen by the primary source, thus the secondary sources become acoustic absorbers. A description and parametric analysis of the process of field cancellation from the frequency domain viewpoint is presented in Chapter 3. The analysis is based on two point sources, of stationary spectral contents, in a three-dimensional non-dispersive propagation medium. The general conditions of cancellation are established. These are then mapped on to the geometrical set-up of sources and sensors which leads to an original geometric analysis of cancellation in terms of the source-related and geometry-related parameters.

The choice of a control structure and design of the controller are the main issues of concern in Chapters 4 and 5. A description of ANC structures and design procedure for the controller are given in Chapter 4. The controller transfer function is derived analytically on the basis of achieving optimum cancellation at an observation point in the medium. An analysis of the system, employing the required controller, leading to results of geometry-related practical limitations in controller implementation is given. A stability analysis of the system in relation to system geometry is presented in Chapter 5. These provide a novel design methodology for ANC systems compliant with control engineering principles.

In a stationary environment, that is, when the noise spectrum is stationary or slowly varying and the system geometry is fixed, the realization of controller transfer function can be achieved either as a digital or analogue filter with fixed transfer characteristics. The procedure of implementing the required controller on a digital signal processor is presented in Chapter 6. A procedure for the measurement and realization of controller characteristics is given. This leads to a controller applicable for cancellation of stationary broadband noise emanating from a compact source. The performance of the ANC system employing such a controller is investigated and discussed through a set of experiments with synthetic and practical noise sources.

A vast number of noise sources encountered in practice are of a time-varying nature. Moreover, the geometric set-up of sources and sensors may be subject to changes. Under such circumstances, the system should be designed within an adaptive control framework so that the required level of performance is maintained. Subsequent chapters of the book concentrate on a stochastic description of ANC systems with two point sources in a three-dimensional non-dispersive propagation medium.

Chapter 7 presents an introduction to adaptive control emphasizing self-tuning control. A description and derivation of a recursive least squares

(RLS) identification algorithm is given. The procedure of implementing the algorithm in real time on a digital signal processor is described and discussed.

Chapter 8 presents the design and implementation of a self-tuning ANC system. It combines the methods of system identification developed in Chapter 7 and the controller design procedure developed in Chapter 4 leading to the design of an adaptive controller. The development of the concept of system identification within an ANC system leads to a procedure for the on-line identification and subsequent implementation of the required controller transfer function for optimum cancellation of broadband noise emanating from a compact source in three dimensions. A novel development of the concept of self-tuning control within a general ANC system framework is given. Moreover, a supervisory level of control is provided based on a performance criterion that activates self-tuning action. The performance of the system is investigated and discussed.

Chapter 9 concludes the book by presenting an overall summary and a discussion of the future prospects of active noise control.

2

DEVELOPMENT OF ACTIVE NOISE CONTROL SYSTEMS

2.1 Introduction

Active noise cancellation employs the principle of acoustic interference and/or absorption. A general theory of interference of waves in three dimensions was given by Huygens in 1690, now known as the Huygens principle, which is described as follows.

Given a (primary) source S_p radiating a (sound) field F_p into a space V; given, further a surface Z_o surrounding S_p and dividing V into two parts V_1, inside Z_o, and V_2, outside Z_o (Fig. 2.1(a)); then a set of (secondary) sources S_H exists which, if properly distributed over Z_o, will radiate a field F_H that is zero in V_1 and equal to F_p in V_2 (Fig. 2.1(b)). Thus, the secondary 'Huygens' sources can be substituted for the primary source with no difference in radiated field in V_2. Consider now the set of secondary sources S_s obtained simply by reversing the sign (phase) of the sources S_H. Owing to linearity, these sources will radiate a field $F_s = -F_H$ that is equal to $-F_p$ in V_2. If we now let the sources S_p and S_s interfere, the resulting field $F_c = F_p + F_s$ (confined field) will be equal to $F_p + 0$ in V_1 and $F_p + (-F_p) = 0$ in V_2 (Fig. 2.1(c)).

The prospect of active noise control (ANC) has intrigued and frustrated, scientists and engineers in many countries for over fifty years. The basic idea is relatively simple; however, the limitations of the available electronic technology have restricted practical implementation to a few simple applications. With the development of fast digital signal processing devices, capable of processing signals within the audio-frequency range, an exciting opportunity now exists to make ANC a practical reality. This places the design of ANC systems firmly within the realm of electronic engineering. In particular, the systems currently being developed employ the analysis skills of control engineers and the design expertise of electronic engineers. Active noise control is, potentially, a significant area of application for control and electronic technology. Moreover, with the development of solid-state electronics making sophisticated signal processing realizable and inexpensive, a tradeoff between rigorous acoustic analysis and control system design is possible.

Although the first theoretical proposals on ANC were reported as early as 1933, the area is still under development, and, apart from a few special cases,

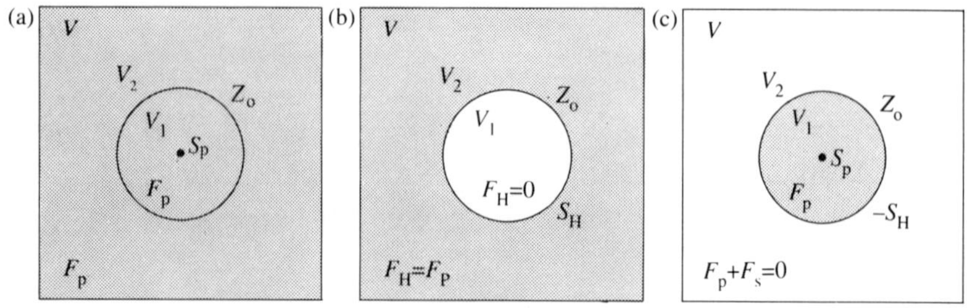

Fig. 2.1. Principle of wave interference: (a) field of primary source; (b) field of secondary source; (c) interference

practically successful ANC systems have still to find wide applications. This chapter presents a retrospective review of ANC systems from the first proposals to the present revival in interest. Applications areas are discussed from an acoustics and control engineering point of view. A consideration of ANC development within deterministic and stochastic descriptions of problems is given.

2.2 Historical review

Active noise control was one of the earliest applications of electronics to the control of physical systems. Lueg filed for a patent in Germany in 1933 and in the USA in 1934 and was granted US Patent No. 2 043 416 in 1936 (Lueg 1936). In his patent, Lueg has made use of the two basic principles of realizing an ANC system: interference and absorption. The interference principle results from the arbitrary physical mixing of acoustic waves leading to constructive and destructive interference which, in turn, cause intensification and weakening of the sound field, respectively. He attempts to manipulate this principle of superposition so that the destructive interference of sound waves could be used to eliminate unwanted noise. He introduced the concept of active attenuation of sound by using artificially generated acoustic waves mixed with the unwanted sound so that the waves were in anti-phase and destructive interference resulted by design. Lueg also uses the absorption principle by synchronizing the movements of a loudspeaker's diaphragm in anti-phase to the unwanted noise so that the noise energy is absorbed by the loudspeaker. Lueg's proposals for obtaining destructive interference and sound energy absorption are shown in Fig. 2.2.

In Fig. 2.2(a) Lueg shows the problem of cancelling sound in a duct. Here acoustic noise propagates along the duct T through point A. The microphone

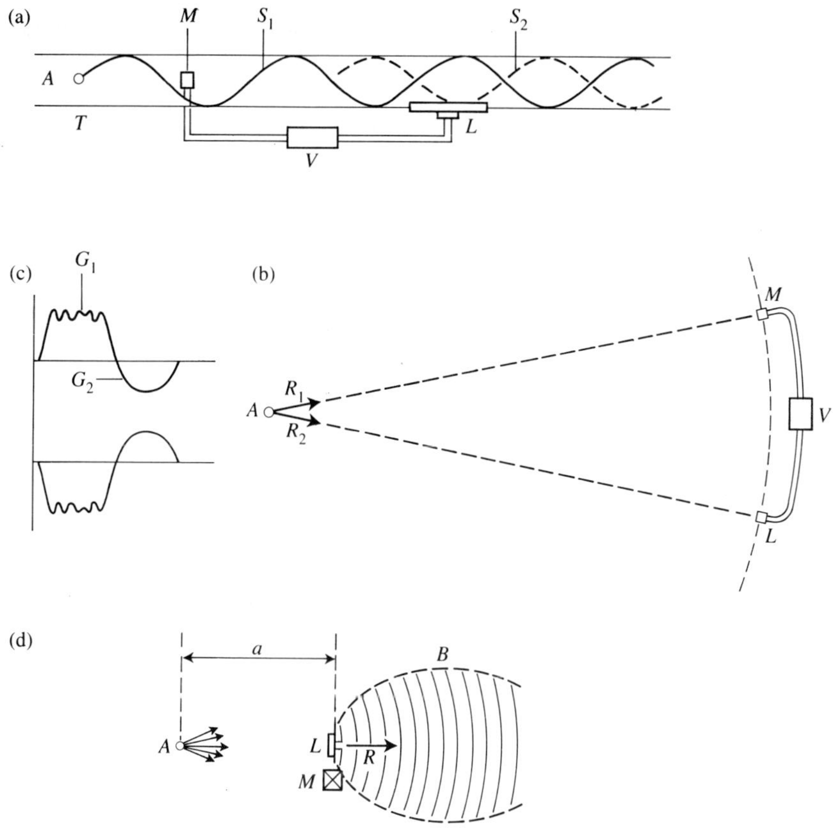

Fig. 2.2. Illustrations from Lueg's patent

M detects the sound and converts it to an electrical signal. The electrical signal passes through amplifier V and then to loudspeaker L. Lueg shows a single frequency in the duct, the phase reversal of which is accomplished by considering the electronic system V as a transmission line with a given time delay. The length of the line is adjusted to give the necessary time delay that results in 180° phase shift relative to the sound wave detected at the microphone so that cancellation of a wave of that particular frequency results.

In Fig. 2.2(b) Lueg shows the problem of cancelling sound oscillations in a limited area near the loudspeaker. Here, the movement of the diaphragm of the loudspeaker is synchronized in anti-phase with the noise arriving at position L. The sound energy is absorbed by the loudspeaker, leading to a zone of sound attenuation around the loudspeaker. The third proposal made by Lueg is shown in Fig. 2.2(d) where he describes the process of noise attenuation in an open space of a point source located at *A*. The microphone *M* detects the sound wave which, after amplification, is delivered to loudspeaker *L* to produce a cancelling wave in the direction of propagation *R* resulting in a zone of silence *B*. In this case attenuation occurs by a process of destructive interference.

In the illustrations (a), (b), and (d) of Fig. 2.2 both interference and absorption principles have been employed. In Fig. 2.2(a) and (b) the unwanted noise is detected and processed by the controller *V* such that by the time the noise reaches the loudspeaker location the movements of the diaphragm of the loudspeaker are in anti-phase synchronization with the noise so that the loudspeaker acts as an absorber. The fact that the loudspeaker acts as an acoustic energy absorber can be verified experimentally by using two loudspeakers, one as a primary source and one as a variable-phase secondary source, both driven by the same single frequency source. When brought close to one another and adjusted so that they are in anti-phase a reduction in the electrical energy taken by the loudspeakers can be detected. Although small, due to the relative inefficiency of the loudspeakers, the effect can clearly be demonstrated by varying the relative phase until no reduction is detectable. The system illustrated in Fig. 2.2(d) employs the principle of interference. Here the detected noise is delayed in phase by 180° and fed to the loudspeaker which in turn superimposes it with the unwanted noise thereby producing destructive interference. The loudspeaker no longer absorbs the sound energy but, by superimposing the anti-phase noise on the unwanted noise, redistributes the acoustic energy in the open space producing regions of cancellation and reinforcement.

In Fig. 2.2(a), Lueg illustrates the basic physical phenomenon which provides the possibility of ANC for this case:

> 'An acoustic wave with a specific frequency has a relatively much lower speed than an electrical signal of the same frequency. This implies that, while a sound wave is travelling from a point where it is detected to a point where it is to be attenuated, there is enough time available within the electronic circuit to process the signal and activate the control elements, to a greater or lesser degree, depending on the frequency, type of noise, and physical extent of the system'.

Note that this basic phenomenon is effective with the structure shown in Fig. 2.2(a) where the relative distance of the detector from the primary source is less than that of the secondary source relative to the primary source. Moreover, if the noise is of a broadband nature the controller path of the system cannot be considered as a simple transmission line but is required to realize a continuous transfer function that phase shifts each detected component frequency by 180°, relative to the noise.

In applying the above-mentioned physical phenomenon, Lueg shows that within the time interval required for the passage of an acoustic wave from the detection point M to the control point L sound can be detected by the microphone, passed through the amplifier (controller) V and fed to the loudspeaker. In this process if the acoustic wave does not change (stationary process) then it will be cancelled at position L provided the phase difference between the wave emitted by the loudspeaker and the unwanted acoustic wave is an odd multiple of 180°. If the controller path is of a given time delay then cancellation of only a specific frequency will occur. Waves at other frequencies will be less well attenuated or even reinforced, depending on their frequency. In general, to achieve good cancellation the controller is to have the required phase-correction as well as amplitude-correction characteristics. Unfortunately, the electronic technology of the 1930s was not sufficiently advanced to meet the requirements of ANC systems and practical results were not realized.

There does not appear to have been any further development by Lueg, and interest in ANC subsided for about twenty years until Olson introduced his 'electronic sound absorber' (Olson and May 1953). In this paper, and later, Olson (1956, 1961) proposed localized sound reducers for occupants of vehicles and for machine operators, machinery noise control, noise reducing headsets, duct noise reduction, and so on.

Olson's absorber is shown in Fig. 2.3(a). The unwanted noise is detected by the microphone and passed through the amplifier. The amplifier, in turn, drives the loudspeaker so that the sound pressure at the microphone location is reduced. This effectively creates a 'zone of cancellation' in front of the absorber. As might be expected from the advance of technology in the 1950s, Olson's device shows considerable sophistication and ingenuity. The problem of the back wave is taken care of by using the absorbing material. The phase problem is minimized by locating the (detector) microphone close to the loudspeaker and ensuring reasonable linearity in the electronics. In fact, it was then possible to control the phase with a good degree of accuracy over a reasonably broad frequency range. However, the basic feedback structure of the system along with other system errors limits the utility of Olson's device both in terms of broadband noise cancellation and the physical extent of cancellation.

The feedback structure of Olson's device is the greatest contributor, among other factors, in limiting its utility. To achieve good cancellation, the closed loop formed by the detector microphone, amplifier, loudspeaker, and the spatial gap between the loudspeaker and the microphone (Fig. 2.3(a)) is required to have impractically large gain. Increasing the loop gain will cause the system to become unstable thereby limiting the performance of the system.

The performance of Olson's device, as a 'sound pressure reducer' at the detector location, is shown in Fig. 2.3(b). The high-frequency performance of the system is limited by the feedback control structure of the system and the

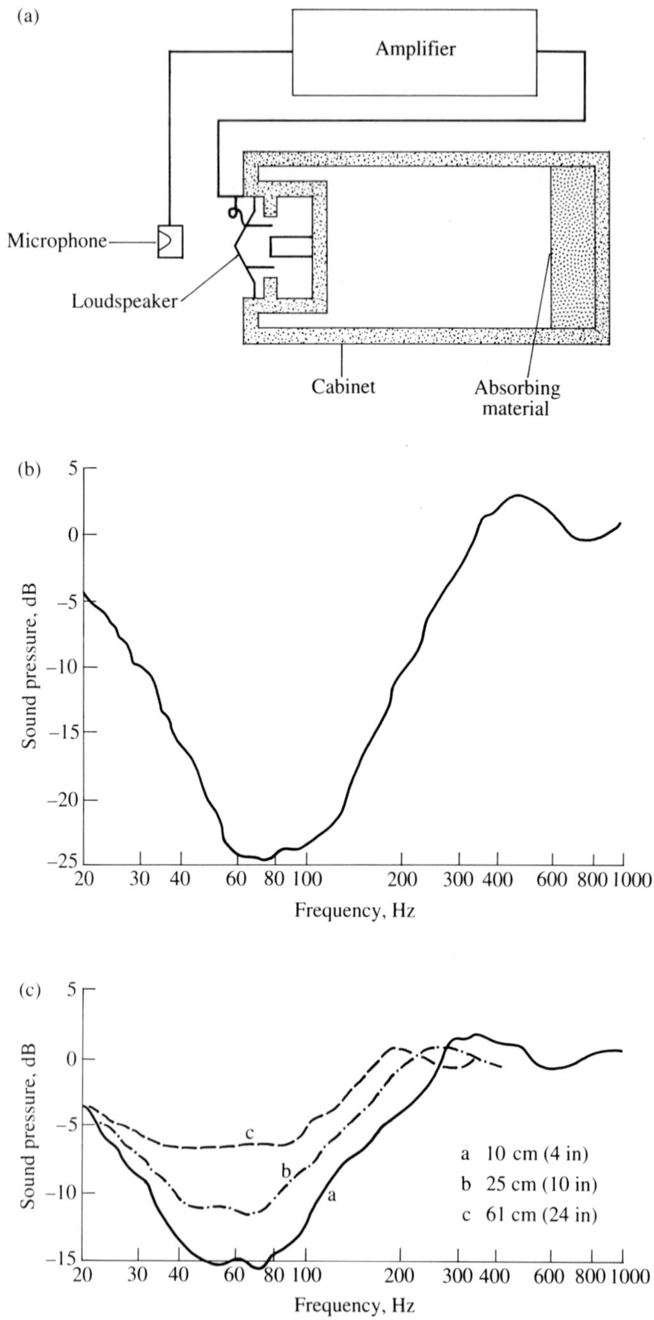

Fig. 2.3. Olson's electronic sound absorber: (a) the device; (b) performance at microphone location; (c) performance at various locations relative to microphone

limited electronic capabilities, whereas the low-frequency performance limitation is due to the response of the loudspeaker in that range. Nevertheless, the performance of the device as measured at the microphone location is promising. Unfortunately, as shown in Fig. 2.3(c), both the amount and frequency range of attenuation decrease for increasing distances from the microphone, making the physical extent of attenuation so limited that at distances of more than about twelve inches from the microphone the device is of no practical use. This is due to the interference pattern set up by the two sources redistributing the sound energy. An area of cancellation is produced around the loudspeaker but larger areas of reinforcement are produced at locations further from the source. There do not seem to be any further developments by Olson; however, his proposal has lately been investigated by others (Bleazey 1962; Chaplin *et al.* 1985; Short 1980; Trinder and Nelson 1983*a,b*).

In the mid-1950s Simshauser and Hawley (1955) proposed the development of an 'active ear defender': a device to reduce ambient noise at the ear by using a headset to generate a sound pressure equal in magnitude and opposite in phase to the noise. This is an area of considerable interest. In a noisy environment where the interference from the ambient noise with telephone conversation, especially at low frequencies, could be at a significant level, conventional ear defenders which use passive techniques of noise reduction are incapable of providing sufficient noise attenuation. It has been proposed that additional attenuation at low frequencies can be obtained by feeding an anti-phase signal to the communications telephone within the ear defender. The device developed by Simshauser and Hawley is a two-channel system consisting of two microphones mounted on either earphone of a conventional military headset, with an amplifier and a phase-shifting network for each channel (Hawley and Simshauser 1961; Simshauser and Hawley 1955; Simshauser *et al.* 1956). Noise is detected by the microphone, passed through the amplifier and phase-shifting network and applied to the earphone. Simshauser and Hawley tested the device in a pure-tone field with proper amplitude and phase adjustments for minimum loudness. The results of their calculations, based on experimental measurements, indicated that such a device would provide an average reduction of 10 dB more than is provided by the earcap alone, in the range of 100 Hz to 1200 Hz. This is currently one of the more successful applications of ANC (Chaplin *et al.* 1987; Dorey *et al.* 1975; Harrison *et al.* 1986; Kim *et al.* 1983; Powell *et al.* 1987; Rodriguez *et al.* 1987; Tichy *et al.* 1987; Twiney *et al.* 1985; Wheeler and Halliday 1981; Wheeler *et al.* 1978).

At about the same time as Olson and Simshauser, Conover (1956, 1957) of General Electric discussed the active control of transformer noise. His work was based on a relatively large, 15 MVA, transformer installation. Unlike Olson and Simshauser he did not confine his work to the laboratory but was rather successful in that he developed his work in the field. Conover's scheme was to place loudspeakers near the transformer's surface and cancel the pressure radiation in the near field. He argued that sound radiates from a transformer due to vibrations of the core, caused principally by magnetostriction, and is transmitted through both the core mountings and the fluid cooling medium to the tank, which in turn couples these vibrations to the surrounding air. This produces a periodic spectrum with harmonics at even multiples of the supply frequency. In a 50 Hz transformer, for instance, vibration frequencies are multiples of 100 Hz and the mode of the tank vibration is, in general, different for each frequency. Furthermore an important characteristic of transformer noise is that the first few harmonics are usually the most important contributors to the sound level. For example the 100, 200, or 300 Hz component is likely to outweigh the combined effect of upper harmonics. Accordingly, a substantial reduction of one or more of the first three components is usually sufficient to achieve a reduction in the noise.

Conover used the 60 Hz supply to the transformer to generate the 120, 240 and 360 Hz harmonics directly. This is a major advantage of this application as no acoustic detection is required and therefore acoustic feedback is eliminated. After adjusting each component in amplitude and phase the signals were recombined, amplified, and fed to a speaker placed at the centre of one of the flat faces of the transformer. Measurements at 100 ft radial distance from the transformer showed a reduction of more than 6 dB within an angular zone of about 11.5° on either side of the line drawn radially perpendicular to the speaker face. Beyond this zone, as shown in Fig. 2.4, reinforcement took place. An important observation made through measurements over time was that the system performance changed and deteriorated as time passed. This is due to the time-varying nature of practical sources of noise. Conover suggested the design of a 'black box' controller to adjust the amplitude and phase of the cancelling signal in accordance with the actual measurements of the noise detected by a microphone at a remote location from the transformer–speaker combination.

Conover was the first to attempt the attenuation of transformer noise by active methods. Since then this has become a classic problem investigated directly or indirectly by Kido (1975), Hesselman (1978), Ross (1978), Jessel and Angevine (Angevine 1981; Jessel and Angevine 1980) and others (Berge *et al.* 1988). Much work has been devoted to the transformer noise because it is potentially so attractive. Transformer hum is annoying and pervasive; it is of economic importance and seems readily solvable by active means because the noise frequencies involved are periodic in nature and precisely related to the supply.

As depicted in Fig. 2.4, Conover's work resulted in a limited ray or beam of cancellation at the expense of intensifying the noise in other areas. It is now clear that the noise of a physical source may be reduced everywhere in the far

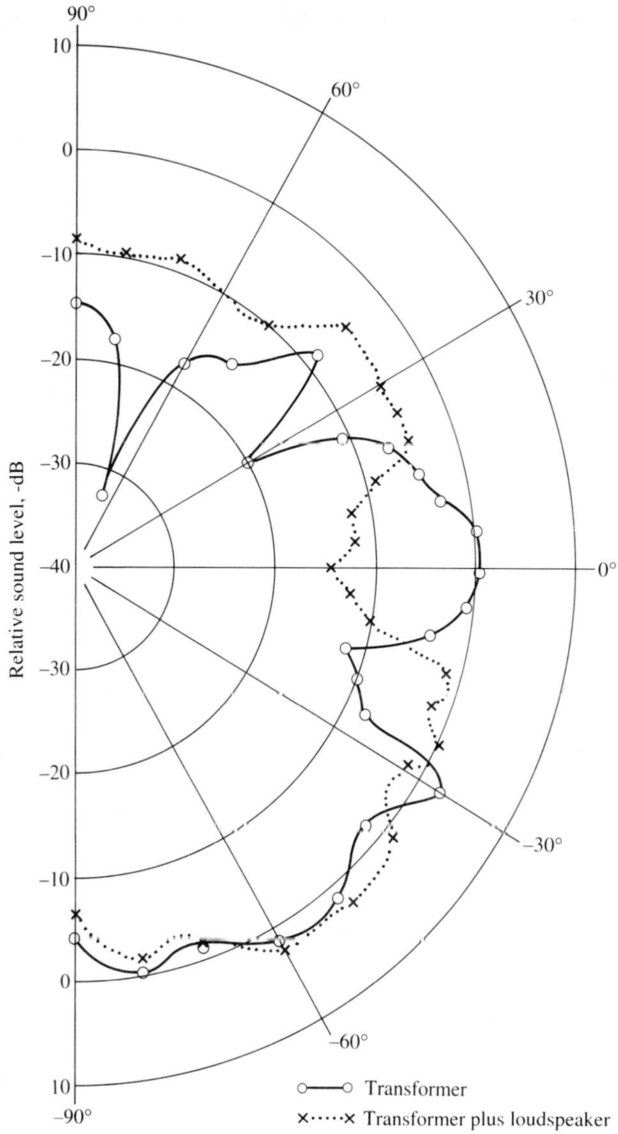

Fig. 2.4. Noise reduction of a 15 MVA transformer, by Conover

field, to some degree, as long as the source is much smaller than the wavelength of the maximum frequency component λ_m of the radiated sound, and the cancelling source and radiating source are located less than $\lambda_m/3$ or $\lambda_m/4$ apart (Hesselman 1978; Warnaka 1982*a*). When the size of the source is much less than the wavelength of the radiated sound, the vibration of the surface is generally in phase. In this case the combined acoustic pressures of the source and the out-of-phase cancelling signal tend to cancel each other near the source. On the other hand, if the physical size of the source is large compared to the wavelength λ_m of the radiated sound the surface no longer moves in a simple uniform phase relationship. The movement is rather in a more complicated pattern since the phase varies over different regions of the source. As a consequence, the radiation pattern of the source is redistributed causing local reduction of the sound pressure without having much effect on the total radiated sound energy. Certain zones are thus created with reduced sound level whereas in other locations the sound is intensified.

The accelerating interest in ANC began in the late 1960s with the publications of Jessel and his co-workers in France, and Kido and his associates in Japan. Jessel's work has been mainly concerned with duct noise (Jessel 1979; Jessel and Mangiante 1972; Jessel and Yamada 1987; Yamada *et al.* 1979). He and his co-workers have also made significant contributions to the theory of active attenuators and general consideration of the field (Jessel 1979, 1981, 1983, 1985; Jessel and Angevine 1980; Jessel and Mangiante 1972). Their contributions have largely been based on the development of Huygens' principle. Kido's work (1975), on the other hand, has been concerned with transformer noise. He and his associates have also made contributions in other applications of ANC (Kido *et al.* 1984; Kim *et al.* 1983, 1984).

It was realized rather early by both Jessel and Kido that the primary advantage of ANC systems is in their ability to attenuate low-frequency noise. This is an area of considerable interest because of the pervasiveness of low-frequency sources and the high cost, large bulk, and relative inefficiency of current passive hardware in low-frequency applications, especially with regard to transmission loss and silencing (Jessel and Angevine, 1980; Kido 1975). Moreover, an advantage in the control of the one-dimensional propagation of duct noise lies in the fact that active duct noise silencers produce no back pressure.

Jessel and others also discovered some of the problems associated with reducing duct noise. Longitudinal duct modes leading to acoustic feedback, due to reflected components, tend to confuse the controllers as to the exact level of the noise itself, since the detector microphone cannot distinguish between the noise and the reflected components. This leads to system instability and/or no noise reduction in some bands of frequency. In order to solve the longitudinal mode problem, so that the detector microphone detects the unwanted noise only, loudspeaker arrays can be used. Two or three loudspeakers used to form an acoustic dipole or an acoustic tripole can be phased to produce acoustic waves travelling in one direction. Two or three microphones in arrays, that is microphone arrays, may be used to obtain directional sensing characteristics which also tend to reduce the longitudinal mode problem.

There has been considerable effort devoted to duct noise since Lueg's invention and primarily since Jessel's first work (Abler and Silcox 1987; Arzamasov *et al.* 1982; Berengier and Roure 1980; Blaser and Chung 1978; Bouc *et al.* 1984; Brewer and Leventhal 1985; Bullmore *et al.* 1985*a,b*, 1987; Burgess 1981; Canevet 1978; Chaplin 1980, 1983; Chaplin and Smith 1979, 1981, 1983; Chaplin *et al.* 1985; Curtis *et al.* 1985; Eghtesadi 1987; Eghtesadi and Chaplin 1987; Eghtesadi and Leventhal 1981*a,b,c*, 1982*a,b*, 1983*a,b*; Eghtesadi *et al.* 1983, 1984, 1985, 1986*a,b*; Elliott and Nelson 1986*b*, 1987; Elliott *et al.* 1987*a*; Enokida *et al.* 1986; Eriksson and Allie 1987; Fedoryuk 1975*b*; Flockton 1987; Ford 1984; F-Williams *et al.* 1985; Gan 1984; Gaudefroy 1981*a,b*; Guicking *et al.* 1983; Heckl 1984; Hong and Eghtesadi 1983; Hong *et al.* 1982, 1983, 1985, 1987; Ivanov 1985; Jessel 1979; Jessel and Mangiante 1972; Jessel and Yamada 1987; Kazakia 1986; Kido *et al.* 1984; Klimov *et al.* 1980; Koers 1986; Kosaka *et al.* 1984; LaFontaine and Shepherd 1983, 1985; Lester and Fuller 1987; Leventhal 1976, 1979, 1980, 1981; Leventhal and Eghtesadi 1979; Martin and Roure 1983; Mazanikov and Tyutekin 1975, 1976, 1978; Mazanikov *et al.* 1977*a,b*; Merkulov and Tyutekin 1980; Michalke 1987; Munjal and Eriksson 1988; Narayana Dutt and Ramakrishna 1978; Nelson *et al.* 1986; Okda *et al.* 1987; Parramore 1974; Poole and Leventhal 1976, 1978; Richarz 1984, 1986; Roebuck 1981; Ross 1981*a,b,c*, 1982*a,b,c,d*; Roure 1985; Roure and Nayroles 1984; Shepherd *et al.* 1984, 1986; Short 1980; Swinbanks 1973, 1976, 1977, 1979, 1981, 1982, 1986, 1987; Takahashi *et al.* 1986, 1987; Taylor 1984; Tichy and Warnaka 1983; Trinder and Nelson 1983*a,b*, Urusovskii 1977*a,b*, 1980, 1981, Vyalyshev *et al.* 1987; Warnaka 1981; Warnaka and Tichy 1980; Warnaka *et al.* 1982; Yamada *et al.* 1979; Zavadskaya and Urusovskii 1976). These have resulted in the development of the schemes depicted in Fig. 2.5. The acoustic monopole system is that originally considered by Lueg. In this system, as the secondary (cancelling) source radiates upstream and downstream, a standing wave is produced upstream which interferes with the detection of the unwanted noise. This makes the position of the microphone extremely sensitive to noise reduction. The acoustic tripole and acoustic dipole were developed by Jessel and his co-workers (Jessel and Mangiante 1972) and Swinbanks (1973), respectively. They attempt to provide a cancelling signal in the duct that propagates only in the downstream direction. The Chelsea system, developed by Leventhal (1976), is formed by two secondary sources with the detector midway between them. The controller is set to nullify the

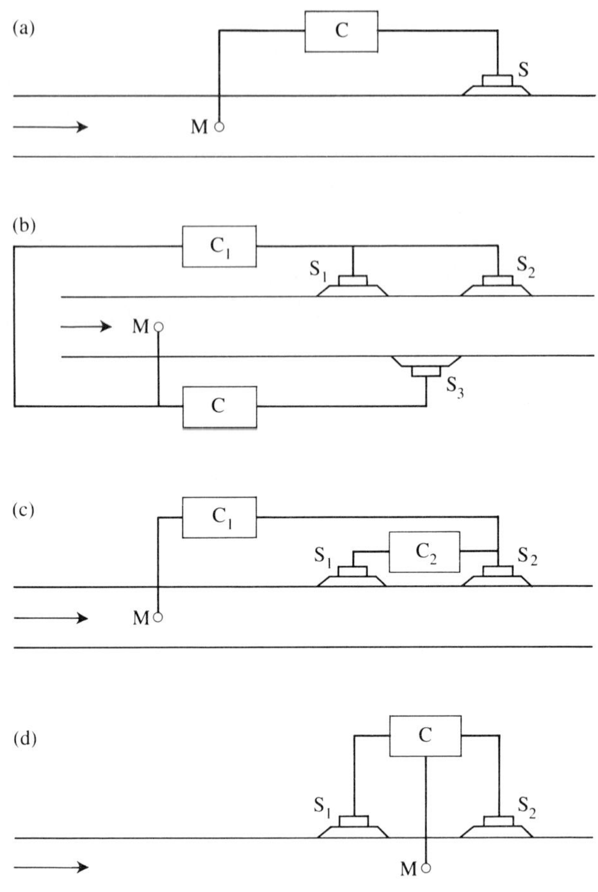

Fig. 2.5. Types of active attenuators for duct noise: (a) monopole (Lueg); (b) tripole (Jessel); (c) dipole (Swinbanks); (d) dual monopole (Chelsea system)

resultant of the secondary sources' waves at the detector location, thus isolating the detector from secondary source radiation. The performance of these systems (Leventhal 1976, 1979) show that they provide noise cancellation of up to 20–25 dB over a narrow band of less than an octave. Maximum cancellation is achieved at only one frequency that is related to the physical spacing of secondary dipole sources (Eghtesadi and Leventhal 1981*b*).

Multiple loudspeaker systems; i.e. dipole and tripole, have essentially been attempts to overcome the problems associated with frequency-dependent controller design. These systems try to isolate the detector microphone from secondary source radiation and/or produce unidirectional radiation, thereby avoiding problems of acoustic feedback and hence instability in the system. However, they have been unable to cope with reflected waves and have geometry-related limitations. The control problem is also much more complex in such systems. The third limitation of these systems is the so-called 'tuning effect' due to the physical spacings of the microphone and loudspeakers relative to each other. By altering these spacings the system is tuned to a different centre frequency, with no significant improvement in the bandwidth of attenuation. For these reasons, the current trend is to use the capabilities of modern electronic systems to implement complex controllers and hence give preference to controller complexity over geometrical complexity (Eghtesadi and Leventhal 1981*a,c*, 1982*a,b*; Eghtesadi *et al.* 1983; Heckl 1984; Hong *et al.* 1983, 1987; Jessel 1985; Leventhal 1981; Leventhal and Eghtesadi 1979; Warnaka and Tichy 1980; Yamada *et al.* 1979). With proper control the monopole is capable of cancelling noise over a broad frequency range. Moreover, the control problem is relatively simple with a single source. However, the task of large attenuation over a broad band primarily lies in the design of the controller employed. What is required, therefore, is a proper ANC system design procedure based on sound analytical conditions for general phase cancellation (Leitch and Tokhi 1987; Leitch *et al.* 1983; Tokhi and Leitch 1988).

The control of noise in three dimensions has received a considerable amount of attention, the greatest attraction being towards transformer noise since Conover's early work (Angevine 1981; Berge *et al.* 1988; Hesselman 1978; Jessel and Angevine 1980; Kempton 1976; Kido 1975; Ross 1978) and more recently towards motor vehicle and machinery noise (Berge 1982, 1983; Bschorr 1972, 1973; Chaplin *et al.* 1985; Chen *et al.* 1971; Corr 1986; Croker 1983; F-Williams 1982; Lamancusa 1987; Leitch and Tokhi 1986; Nadim and Smith 1983; Oswald 1984; Tokhi and Leitch 1989; Trinder *et al.* 1986; Wheeler and Parramore 1981; Wurm and Bergson 1974). However, some other specific applications have also been reported (Bleazey 1962; Dines 1982; Eriksson *et al.* 1988*b*; F-Williams 1984; Kim *et al.* 1984; Kroll 1987; Mazzanti and Piraux 1983; Roberts and Vuillermoz 1987; Ross 1982*e*, 1986; Swanson 1987; Swinbanks 1986; Vandenhout and Sas 1986; Warnaka *et al.* 1983) as well as theoretical contributions towards understanding the general problem (Anfinogentov *et al.* 1980; Brewer and Leventhal 1985; Brown 1985; Bschorr 1971; Bullmore *et al.* 1985*a,b*; Curtis *et al.* 1987; Czarnecki 1970; Darlington and Elliott 1985; Davison and Robinson 1977; Elliott and Nelson 1985*a,b*; Elliott *et al.* 1988; Fedoryuk 1975*a*, 1976, 1979; F-Williams 1984, 1985; Gavrilov *et al.* 1977; Guicking and Rollwage 1984; Guicking *et al.* 1983; Jessel 1979, 1981, 1983; Kempton 1976; Kim *et al.* 1984; Konyaev *et al.* 1977, 1979; Korotaev and Mazanikov 1985; Leitch and Tokhi 1987; Leitch *et al.* 1983; Levine 1980; Mazanikov *et al.* 1980; Mazzanti and Piraux 1983; Nelson *et al.* 1985, 1987*a,b,c*; Piraux and Mazzanti 1985; Piraux and Nayroles 1980; Richarz 1984; Roebuck 1981; Silcox and Elliott 1985; Swinbanks 1986; Tartakovskii 1975; Tohyama and Suzuki 1987; Tokhi and

Leitch 1988; Trinder and Jones 1987; Urusovskii 1981, 1986; Vogt 1976; Vyalyshev *et al.* 1977, 1986; Warnaka 1987; Warnaka *et al.* 1983; Zavadskaya *et al.* 1975*a,b*, 1977). But, until recently, there has been little consideration of general ANC systems in three dimensions (A-Tinoco 1985; Guicking *et al.* 1983; Leitch and Tokhi 1986, 1987; Leitch *et al.* 1983; Mazzanti and Piraux 1983; Piraux and Mazzanti 1985; Piraux and Nayroles 1980; Tokhi and Leitch 1988). A general trend is now seen towards the application of digital signal processing techniques to the active control of noise (A-Tinoco 1985; Chaplin 1980, 1983; Chaplin and Smith 1979, 1981, 1983; Chaplin *et al.* 1979, 1985, 1987; Chaturved and Thomas 1981; Darlington and Elliott 1985; Elliott and Nelson 1986*a*; Elliott *et al.* 1987*b*; Eriksson and Allie 1987; Eriksson *et al.* 1987, 1988*a*; Leitch and Tokhi 1986, 1987; Muller and Pauw 1986; Oswald 1984; Poole *et al.* 1984; Ross 1981*a,b,c*, 1982*a,b,c*; Roure 1985; Roure and Nayroles 1984; Sjosten 1985; Sjosten *et al.* 1986; Smith and Chaplin 1983; Swinbanks 1981; Tokhi and Leitch 1989). This is primarily due to the current advance in digital technology which makes the implementation of digital controllers feasible at a low cost.

2.3 Approaches to design

Looking back at the early developments in ANC, Lueg's initial efforts were unsuccessful, since no papers were published following his patent, and no applications in the 1930s and 1940s are known. In the 1950s Olson, Simshauser and Conover showed considerable interest in ANC. However, their efforts were also unsuccessful. Therefore, interest again subsided in the late 1950s and early 1960s. In retrospect, this has been due to a poor understanding of the problem from the acoustics and systems point of view, as well as the lack of necessary technology to be able to deal with the main problems associated with the design of successful ANC systems. These include:

(1) acoustic response of the medium;
(2) characteristics of transducers and other electronic equipment used;
(3) broadband noise sources;
(4) distributed noise sources;
(5) time-varying noise sources.

An acoustic wave, in propagating through a medium, is affected in amplitude as well as in phase due to the response characteristics of the medium. Thus, each component frequency of the noise emitted undergoes an amplitude and a phase change as it reaches the point where it is detected (measured). Moreover, acoustic feedback and reflected waves have a great effect on the performance of an ANC system. These factors, along with the geometric arrangement of an ANC system, have a major role in the stability of the system.

Consider a uniformly pulsating source placed in a linear propagation medium emitting sound waves in all directions. If there are no boundaries or obstacles to reflect sound waves back towards the source then the sound waves are attenuated in accordance with the inverse square law. This results in a sound level in the medium that decreases for increasing distances from the source. The major factors affecting the performance of an ANC system in such a case are: (a) characteristics of the source, (b) the acoustic properties of the medium, (c) the geometrical arrangement of the system, and (d) the characteristics of the electronic equipment used. However, if the source emits waves into an enclosure then the performance of an ANC system, besides the above factors, is also affected by the acoustic response of the enclosure.

A sound source emitting broadband noise into an enclosure will lead to fluctuations in the sound pressure (resonances). Here the locations of the maxima and minima, with respect to each other, of the sound pressure depend on the size and geometrical shape of the enclosure: for a large, irregular enclosure the maxima and minima lie much closer together in position than for either a small or a rectangular enclosure. Moreover, few room resonances will occur due to low-frequency noise while at high frequencies there will be many resonances in any given band of frequencies. Thus, high-frequency sound waves whose average wavelength is small compared with the dimensions of the enclosure will result in a large number of standing waves. As ANC is intended for the reduction of low-frequency sound there will be a limited number of resonances to consider. Furthermore, a common assumption is that the unwanted (primary) source of noise is considered to be compact (with respect to the frequencies of interest) and that the cancelling (secondary) source of noise is physically close to the primary source, again with respect to the frequencies of interest.

The characteristics of transducers used in an ANC system form another factor in affecting the performance of the system. These characteristics are usually frequency dependent and hence affect each frequency component of a measured acoustic wave. Moreover, the physical dimensions of transducers, say microphones and loudspeakers, with respect to the frequencies of interest play a major role in the performance of an ANC system. Therefore, the selection of transducers for application in ANC systems needs careful consideration (Bschorr 1986; Edwards and Dekker 1985; Fry and Fryer 1985; Glendinning *et al.* 1985; Poole and Warnaka 1987).

It is noted in most papers on ANC that little discussion is devoted to the electronic systems used. Also, frequency domain and control system concepts are only rarely used in the analysis and design of ANC systems (Leitch and Tokhi 1986, 1987; Leitch *et al.* 1983; Ross 1981*a*,*b*,*c*, 1982*a*,*b*,*c*; Tokhi and Leitch 1988, 1989). Similarly, little is found about geometrical considerations in the literature. Details of best arrangements are not given. Analytical approaches, on the other hand, which have treated the problem mathematically, have always been extremely general because of the highly complicated nature of the problem. Nevertheless, the physical geometry has been shown to be extremely important in achieving large attenuation (Leitch and Tokhi 1987; Shepherd *et al.* 1984; Tokhi and Leitch 1988; Warnaka and Tichy 1980). A high degree of attenuation can be expected to be obtained only by geometries which favour proper mixing of the cancelling field with the unwanted noise field.

The noise emitted by practical sources is generally of a broadband nature. Therefore, ANC systems are required to be able to cope with such broadband noise. An ANC system employing an amplifier (which is equivalent to a constant-length transmission line) as a controller can provide maximum cancellation of noise at a single frequency, with cancellation decreasing on either side of this frequency and eventually producing reinforcement. Because of the broadband nature of acoustic sources the controller is required to realize frequency-dependent continuous transfer characteristics. These characteristics are found to be dependent on a number of frequency-dependent items of the system, such as the transfer characteristics of the source, transducers, and other electronic equipment, and the propagation medium (Leitch and Tokhi 1986, 1987; Ross 1981*a,c*, 1982*a,b,c*; Tokhi and Leitch 1988). Such a dependence makes it possible to identify and realize the required transfer characteristics as a controller.

A feature of many noise sources is that they are not compact: noise is not emitted by one small part of the source but is distributed arbitrarily over the entire surface of the source. A distributed noise source is equivalent to a number of compact noise sources (multiple sources), distributed around the source surface. In such a situation the control problem is much more complicated than in the case of a single compact source. Here, a single secondary source is not sufficient to attenuate the unwanted noise but rather a number of them, depending on geometry limitations, are required. Ideally, to achieve cancellation all over the medium surrounding the source, an infinite number of cancelling sources will be required. In practice, the number of cancelling sources is limited by their physical dimensions. However, a reasonable finite number of cancelling sources can be realized so that a required set of performance specifications are met. This will depend on the level of system complexity, level of noise attenuation required within a limited region of the medium, and hardware/software implementation.

Application of acoustic theory and modern control concepts to the control of transformer noise has revealed that a significant amount of reduction (10–30 dB) in the fundamental frequency component can be obtained everywhere in the far field provided the physical source is smaller than the wavelength of the radiated sound, and the cancelling source is located near the radiating source. When the physical source is larger than the wavelength of the radiated sound and/or for harmonics of the fundamental frequency component, only certain limited zones of cancellation can be produced. The size of these zones can be increased by the addition of more detectors and loudspeakers. This redistributes the sound energy in the medium, thereby dividing the medium into zones of cancellation and sound intensification. The situation can be improved by matching an array of detectors and cancelling sources with the subdivided space so that one element of the array coincides with each subdivision. This method has clear limitations in complexity and cost as well as design. The addition of each element to the array improves noise attenuation in the sense that more zones of cancellation are created. Thus a new sound energy distribution pattern is formed with new zones of cancellation and sound intensification. This means that the problem of eliminating the zones of sound intensification will still remain unsolved.

A further factor affecting the success of ANC systems is that the characteristics of many practical sources vary with operating conditions and hence time. This is also the case with the transducers and other electronic equipment used. Alteration of the geometric arrangement of an ANC system, on the other hand, has a direct effect on the performance of the system because the controller transfer function is found to be directly related to the system geometry.

The time-varying phenomenon forms an important dimension in ANC which the designer has to consider. Such a consideration is an essential requirement in many applications, and is the crucial difference between laboratory success and practical realization. Here a controller with fixed frequency-dependent transfer characteristics is of limited use because of the continual adjustment required in the system as the noise pattern of the source varies from time to time. One approach to this problem is to use the concepts of adaptive control so that the transfer characteristics of the controller are adjusted in accordance with variations in the system. Through his experiments of reducing transformer noise, Conover was the first to realize the need for a 'black box' controller that would adjust the cancelling signal in accordance with information gathered at a remote distance from the transformer (Conover 1956). Later it has been realized by numerous authors that an essential requirement for an ANC system to be practically successful is that it be adaptive (A-Tinoco 1985; Burgess 1981; Chaplin 1980; Darlington and Elliott 1985; Elliott and Nelson 1986*a*; Elliott *et al.* 1987*b*; Leitch and Tokhi 1986, 1987; Ross 1982*a,b,c*). Implementing an adaptive control algorithm within the ANC system will allow the system to adjust the controller characteristics in accordance with changes in the noise pattern of the source.

Chaplin and his associates (Chaplin 1980, 1983; Chaplin and Smith 1979, 1981, 1983; Chaplin *et al.* 1979, 1985, 1987; Corr 1986; Croker 1983; Nadim and Smith 1983; Smith and Chaplin 1983) have reported some success with an adaptive scheme based on trial and error waveform generation. This method relies on the source noise being periodic. The method has been tested with rotary machines such as motor vehicles, and fans where the noise spectrum is harmonically related to the fundamental (engine firing) frequency. The engine firing frequency provides a clock signal to the algorithm. The algorithm starts adjusting the controller parameters in a particular direction. If this leads to an improvement in performance the adjustment continues in the same sense. However, if, at any stage, the performance deteriorates the controller parameters are adjusted in the opposite sense. There is no doubt that the implementation of an adaptive controller based on a stochastic parameter estimation algorithm is a potentially attractive solution to reducing broadband noise under time-varying conditions. Recently, some theoretical and practical investigations have been reported in this area (Darlington and Elliott 1985; Dines 1982; Elliott and Nelson 1986*a*; Elliott *et al.* 1987*b*; Eriksson *et al.* 1987, 1988*a*; Roure and Nayroles 1984; Sjosten 1985; Sjosten *et al.* 1986; Tokhi and Leitch 1989). Among these the scheme developed by Eriksson and his co-workers relies on the development of a model of the source. The scheme reported by Elliott and his co-workers is based on minimization (in the least squares sense) of sound level at discrete locations in the medium. The self-tuning control scheme developed by Tokhi and Leitch (detailed later in this book) is based on the optimum cancellation of noise.

Adaptive ANC schemes rely on either a non-recursive or recursive realization of the controller. Practical investigations in digital signal processing have shown that with a non-recursive structure a high order and hence large number of parameters are required for the controller to approximate the required transfer characteristics. Moreover, in ANC applications a change in the geometric arrangement of the system leads to a substantial change in the order of such a controller. These can lead to limitations in data memory and to problems of increased computation and hence processing time. This further limits the frequency band and physical extent of cancellation, making the algorithm suitable for applications where noise cancellation is required at narrow bands, near the source, and within limited areas. The advantage of this structure, however, is that the controller is always stable. With a recursive realization structure, on the other hand, there are relatively few controller parameters involved. Therefore, the controller characteristics can be updated for broadband noise cancellation and over large areas around the source. However, depending on design methodology and the accuracy of convergence of the algorithm, the controller may become unstable leading to an unstable system.

Advances in technology which began prior to the mid-1960s, and which are continuing, make the active control of noise much more attractive and feasible than it has ever been before. The areas of technology where the most important applicable progress has occurred are: advanced control systems technology, which includes the development of adaptive algorithms; improved understanding of the physics of acoustic systems; and the development of solid-state electronics, making sophisticated control realizable and inexpensive.

The field of ANC is currently attracting considerable worldwide attention. There are programmes continuing in many countries including Britain, France, Germany, Japan, Poland, Russia, and the USA. There is no doubt that the rapid development of this field will result in successful applications in ANC in the years to come. This is because of the advantages and capabilities of ANC systems as compared with passive methods, namely:

(1) considerably improved low-frequency performance;
(2) reduction in hardware costs for many applications;
(3) great reduction of size and weight;
(4) zero, or very low, back pressure.

The old concept of active noise attenuation is becoming a reality outside the laboratory because new technological developments are providing the opportunity for the advantages to be realized. Advances are currently being made towards developing commercially available hybrid active–passive noise attenuators in which the low-frequency attenuation, below 500 Hz, is provided by an active system, while the high-frequency attenuation is provided by passive hardware.

2.4 Summary

Active noise control is realized by using either one or both of the basic principles of interference and absorption. The interference principle consists of the mixing of acoustic waves resulting in constructive and destructive interference which, in turn, causes intensification and weakening of the sound field, respectively. The absorption principle on the other hand, results by synchronizing the secondary source, say a loudspeaker's diaphragm, in antiphase to the unwanted noise so that the noise energy is absorbed by the loudspeaker.

The concept of ANC has intrigued scientists and engineers in many countries for over half a century. The early efforts, prior to the mid-1960s, were unsuccessful. This was due to a poor understanding of the problem from the acoustics and systems points of view, as well as the limitations in technology to be able to deal with the main problems associated with the design of ANC systems.

Advances in technology which began by the mid-1960s, and are continuing, make ANC much more feasible and attractive. These include advanced control systems technology (including the development of adaptive algorithms), improved understanding of the physics of acoustic systems, and the development of solid-state electronics.

The main part of an ANC system which needs to be given greater consideration is a design procedure for the controller. An ANC system employing an amplifier (which is equivalent to a constant-length transmission line) as a controller can provide maximum cancellation of the unwanted noise at only one frequency, with cancellation decreasing on either side of this frequency and eventually producing reinforcement. Because of the broadband nature of acoustic sources, the controller is required to realize continuous frequency-dependent transfer characteristics so that it is capable of producing the mirror image of every detected frequency component of the unwanted noise. In the past, because of the complexity involved in implementing such a controller, geometrical complexity, by using a number of secondary sources, was preferred. However, with the capabilities of current electronic technology, especially digital electronics, it is possible to implement such a controller either as a fixed or as an adaptive digital controller, whichever is required by the application, on a special-purpose microprocessor.

The capability of today's electronic technology provides the opportunity to develop sophisticated ANC systems that can 'self-tune' to the required controller characteristics, thereby minimizing the need for extensive analysis. Such systems will also allow the system to 'track' the variation in spectral content of real sources such that stable cancellation is maintained over a range of operating conditions. This adaptive capability is essential to the realization of successful ANC systems for real industrial applications.

2.5 References

Abler, S.B. and Silcox, R.J. (1987). Experimental evaluation of active noise control in a thin cylindrical shell. *Proceedings National Conference on Noise Control Engineering*, Pennsylvania, pp. 341–6.

Anfinogentov, V.I., Lyubashevskii, G.S., Tartakovskii, B.D., Filippov, V.E., and Choni, Yu.I. (1980). Control algorithms for active vibroacoustic suppression systems. *Soviet Physics Acoustics*, English translation, **26**, (4), 340–1.

Angevine, O.L. (1981). Active acoustic attenuation of electric transformer noise. *Proceedings of Inter-noise 81: International Conference on Noise Control Engineering*, Amsterdam, **1**, 303–6.

Arzamasov, S.N., Malakhov, A.N., and Mal'tsev, A.A. (1982). Adaptive system for the active suppression of noise fields in a multimode waveguide. *Soviet Physics Acoustics*, English translation, **28**, (5), 346–8.

A-Tinoco, A.M. (1985). Adaptive algorithms for the active attenuation of acoustic noise. Unpublished Ph.D. thesis. Heriot–Watt University, Edinburgh.

Berengier, M. and Roure, A. (1980). Broad-band active sound absorption in a duct carrying uniformly flowing fluid. *Journal of Sound and Vibration*, **68**, (3), 437–49.

BERGE, T.S. (1982). A feasibility study of active noise cancellation of low frequency sound inside vehicle cabs. Unpublished M.Sc. Dissertation, University of Southampton.

BERGE, T.S. (1983). Active noise cancellation of low frequency sound inside vehicle cabs. *Proceedings of Inter-noise 83: International Conference on Noise Control Engineering*, Edinburgh, **I**, 457–60.

BERGE, T., PETTERSEN, O.KR.O., and SORSDALL, S. (1988). Active cancellation of transformer noise: field measurements. *Applied Acoustics*, **23**, (4), 309–20.

BLASER, D.A. and CHUNG, J.Y. (1978). A transfer function technique for determining the acoustic characteristics of duct systems with flow. *Proceedings of Inter-noise 78: International Conference on Noise Control Engineering*, San Francisco, pp. 901–8.

BLEAZEY, J.C. (1962). Electronic sound absorber. *Journal of Audio Engineering Society*, **10**, (2), 135–9.

BOUC, R., NAYROLES, B., and AGHFIR, M. (1984). Recursive filtering of an acoustic wave in a turbulent flow and active absorption in ducts. *Proceedings of Inter-noise 84: International Conference on Noise Control Engineering*, Honolulu, **I**, 443–6.

BREWER, P.A. and LEVENTHAL, H.G. (1985). Active attenuation in small enclosures. *Proceedings of the Institute of Acoustics*, **7**, Part 2, 111–14.

BROWN, I. (1985). The application of simple source theory to active noise control. *Proceedings of the Institute of Acoustics*, **7**, Part 2, 73–8.

BSCHORR, O. (1971). *Sound shielding by means of sound.* US Patent No. 3 602 331.

BSCHORR, O. (1972). *Noise reduction for propellers.* US Patent No. 3 685 610.

BSCHORR, O. (1973). *Means for the reduction of noise of rotors.* UK Patent No. 1 304 329.

BSCHORR, O. (1986). An integrated microphone/loudspeaker unit for active noise cancellation. *Proceedings of Inter-noise 86: International Conference on Noise Control Engineering*, Cambridge, Mass., **II**, 1447–50.

BULLMORE, A.J., NELSON, P.A., and ELLIOTT, S.J. (1985*a*). Active control of harmonic enclosed sound fields of low modal density; a computer simulation. *Proceedings of the Institute of Acoustics*, **7**, Part 2, 55–64.

BULLMORE, A.J., ELLIOTT, S.J., and NELSON, P.A. (1985*b*). The active minimisation of periodic sound in enclosures; an experimental investigation. *Proceedings of Inter-noise 85: International Conference on Noise Control Engineering*, Munich, **I**, 575–8.

BULLMORE, A.J., NELSON, P.A., CURTIS, A.R.D., and ELLIOTT, S.J. (1987). The active minimization of harmonic enclosed sound fields, Part II: a computer simulation. *Journal of Sound and Vibration*, **117**, (1), 15–33.

BURGESS, J.C. (1981). Active adaptive sound control in a duct: A computer simulation. *Journal of the Acoustical Society of America*, **70**, (3), 715–26.

CANEVET, G. (1978). Active sound absorption in an air-conditioning duct. *Journal of Sound and Vibration*, **58**, (3), 333–45.

CHAPLIN, B. (1980). The cancellation of repetitive noise and vibration. *Proceedings of Inter-noise 80: International Conference on Noise Control Engineering*, Florida, **II**, 699–702.

CHAPLIN, B. (1983). Anti-noise – the Essex breakthrough. *Chartered Mechanical Engineering*, **30**, 4107.

CHAPLIN, G.B.B. and SMITH, R.A. (1979). *Improvements in and relating to 'active' methods for attenuating compressional waves.* UK Patent No. 1 555 760.

CHAPLIN, G.B.B. and SMITH, R.A. (1981). Active control of repetitive noise and vibration. *Proceedings of the Institute of Acoustics*, February, 9–12.

CHAPLIN, G.B.B. and SMITH, R.A. (1983). Waveform synthesis – the Essex solution to repetitive noise and vibration. *Proceedings of Inter-noise 83: International Conference on Noise Control Engineering*, Edinburgh, **I**, 399–402.

CHAPLIN, G.B.B., SMITH, R.A., and BEARCROFT, R.G. (1979). *Active attenuation of recurring sounds.* US Patent No. 4 153 815.

CHAPLIN, G.B.B., JONES, A., and JONES, O. (1985). *Methods and apparatus for low frequency active attenuation.* US Patent No. 4 527 282.

CHAPLIN, G.B.B., SMITH, R.A., and BRAMER, T.P.C. (1987). *Methods and apparatus for reducing repetitive noise entering the ear.* US Patent No. 4 654 871.

CHATURVED, G.K. and THOMAS, D.W. (1981). Adaptive noise cancelling and condition monitoring. *Journal of Sound and Vibration*, **76**, (3), 391–405.

CHEN, Y.R., HUANG, B.K., and SUGGS, S.W. (1971). Noise control by acoustic interference. *Transactions of ASME*, **3**, 424–32.

CONOVER, W.B. (1956). Fighting noise with noise. *Noise Control*, **92**, 78–82, 92.

CONOVER, W.B. (1957). *Noise reducing system for transformers.* US Patent No. 2 776 020.

CORR, H.F.J. (1986). Active control of exhaust noise from a commercial vehicle. *Proceedings of the Institute of Acoustics*, **8**, Part 1, 159–64.

CROKER, M.D. (1983). The active control of internal combustion engine exhaust noise. *Proceedings of Inter-noise 83: International Conference on Noise Control Engineering*, Edinburgh, **I**, 451–5.

CURTIS, A.R.D., NELSON, P.A., and ELLIOTT, S.J. (1985). Active control of one-dimensional enclosed sound fields. *Proceedings of Inter-noise 85: International Conference on Noise Control Engineering*, Munich, **I**, 579–82.

CURTIS, A.R.D., NELSON, P.A., ELLIOTT, S.J., and BULLMORE, A.J. (1987). Active suppression of acoustic resonance. *Journal of the Acoustical Society of America*, **81**, (3), 624–31.

CZARNECKI, S. (1970). Investigation of sound transmission properties of a medium resulting from sound wave compensation caused by other sound sources. *Journal of Sound and Vibration*, **11**, (2), 225–33.

DARLINGTON, P. and ELLIOTT, S.J. (1985). Adaptive control of periodic disturbances in resonant systems. *Proceedings of the Institute of Acoustics*, **7**, Part 2, 87–94.

DAVISON, A.R. Jr. and ROBINSON, T.G.F. (1977). *Noise cancellation apparatus.* US Patent No. 4 025 724.

DINES, P.J. (1982). Comparison of least squares estimation and impulse response techniques for active control of flame noise. *Proceedings of the Institute of Acoustics*, November, F2.1–F2.4.

DOREY, A.P., PELC, S.F., and WATSON, P.R. (1975). An active noise reduction system for use with ear defenders. *Proceedings of 8th International Aerospace Symposium*, Cranfield, **19**, (5), 24–7.

EDWARDS, J.W. and DEKKER, N.M. (1985). Experimental verification of a low frequency loudspeaker model for an engineering application. *Proceedings of the Institute of Acoustics*, **7**, Part 3, 267–75.

EGHTESADI, KH. (1987). Active absorption of one-dimensional sound field by *n*-source attenuator arrays. *Proceedings of National Conference on Noise Control Engineering*, Pennsylvania, pp. 353–8.

EGHTESADI, KH. and CHAPLIN, G.B.B. (1987). The cancellation of repetitive noise and vibration by active methods. *Proceedings of National Conference on Noise Control Engineering*, Pennsylvania, pp. 347–52.

EGHTESADI, KH. and LEVENTHAL, H.G. (1981*a*). Monopole system for active attenuation of noise in ducts. *Proceedings of the Institute of Acoustics*, February, 13–16.

EGHTESADI, KH. and LEVENTHAL, H.G. (1981*b*). Active attenuation of sound: the Chelsea dipole. *Journal of Sound and Vibration*, **75**, (1), 127–34.

EGHTESADI, KH. and LEVENTHAL, H.G. (1981*c*). Comparison of active attenuators of noise in ducts. *Acoustics Letters*, **4**, (10), 204–9.

EGHTESADI, KH. and LEVENTHAL, H.G. (1982*a*). Active attenuation of noise – the monopole system. *Journal of the Acoustical Society of America*, **71**, (3), 608–11.

EGHTESADI, KH. and LEVENTHAL, H.G. (1982*b*). Monopole active attenuators with non-ideal loudspeakers. *Acoustics Letters*, **5**, (11), 169–74.

EGHTESADI, KH. and LEVENTHAL, H.G. (1983*a*). The effects of non-ideal elements and geometry on the performance of the Chelsea dipole active attenuators. *Journal of Sound and Vibration*, **91**, (1), 1–10.

EGHTESADI, KH. and LEVENTHAL, H.G. (1983*b*). A study of *n*-source active attenuator arrays for noise in ducts. *Journal of Sound and Vibration*, **91**, (1), 11–19.

EGHTESADI, KH., HONG, W.K.W., and LEVENTHAL, H.G. (1983). The tight-coupled monopole active attenuator in a duct. *Noise Control Engineering Journal*, **20**, (1), 16–20.

EGHTESADI, KH., HONG, W.K.W., and LEVENTHAL, H.G. (1984). Economics of active attenuation of noise in ducts. *Proceedings of Inter-noise 84: International Conference on Noise Control Engineering*, Honolulu, **I**, 447–52.

EGHTESADI, KH., HONG, W.K.W., and LEVENTHAL, H.G. (1985). Applications of active attenuators in controlling low-frequency noise. *Proceedings of the National Conference on Noise Control Engineering*, Ohio, pp. 193–8.

EGHTESADI, KH., HONG, W.K.W., and LEVENTHAL, H.G. (1986*a*). Energy conservation by active noise attenuation in ducts. *Noise Control Engineering Journal*, **27**, (3), 90–4.

EGHTESADI, KH., HONG, W.K.W., and LEVENTHAL, H.G. (1986*b*). Evaluation of active noise attenuator systems. *Proceedings of Inter-noise 86: International Conference on Noise Control Engineering*, Cambridge, Mass., **I**, 577–82.

ELLIOTT, S.J. and NELSON, P.A. (1985*a*). Error surfaces in active noise control. *Proceedings of the Institute of Acoustics*, **7**, Part 2, 65–72.

ELLIOTT, S.J. and NELSON, P.A. (1985*b*). The active minimisation of sound fields. *Proceedings of Inter-noise 85: International Conference on Noise Control Engineering*, Munich, **I**, 583–6.

ELLIOTT, S.J. and NELSON, P.A. (1986*a*). An adaptive algorithm for multichannel active control. *Proceedings of the Institute of Acoustics*, **8**, Part 1, 135–47.

ELLIOTT, S.J. and NELSON, P.A. (1986*b*). The implications of causality in active control. *Proceedings of Inter-noise 86: International Conference on Noise Control Engineering*, Cambridge, Mass., **I**, 583–8.

ELLIOTT, S.J. and NELSON, P.A. (1987). The active control of enclosed sound fields. *Proceedings of National Conference on Noise Control Engineering*, Pennsylvania, pp. 359–64.

ELLIOTT, S.J., CURTIS, A.R.D., BULLMORE, A.J., and NELSON, P.A. (1987*a*). The active minimization of harmonic enclosed sound fields, Part III: experimental verification. *Journal of Sound and Vibration*, **117**, (1), 35–58.

ELLIOTT, S.J., SOTHERS, I.M., and NELSON, P.A. (1987*b*). A multiple error LMS algorithm and its application to the active control of sound and vibration. *IEEE Transactions on Acoustics, Speech, and Signal Processing*, **35**, (10), 1423–34.

ELLIOTT, S.J., JOSEPH, P., BULLMORE, A.J., and NELSON, P.A. (1988). Active cancellation at a point in a pure tone sound field. *Journal of Sound and Vibration*, **120**, (1), 183–9.

ENOKIDA, T., HAMADA, H., MIURA, T., TAKAHASHI, M., KURIBAYASHI, T., and ASAMI, K. (1986). Design of electric sound cancellation system for air conditioning duct. *Proceedings of Inter-noise 86: International Conference on Noise Control Engineering*, Cambridge, Mass., **I**, 589–94.

ERIKSSON, L.J. and ALLIE, M.C. (1987). A digital sound control system for use in turbulent flows. *Proceedings of National Conference on Noise Control Engineering*, Pennsylvania, pp. 365–70.

ERIKSSON, L.J., ALLIE, M.C., and GREINER, R.A. (1987). The selection and application of an IIR adaptive filter for use in active sound attenuation. *IEEE Transactions on Acoustics, Speech, and Signal Processing*, **35**, (4), 433–7.

ERIKSSON, L.J., ALLIE, M.C., BREMIGAN, C.D., and GREINER, R.A. (1988*a*). Active noise control using adaptive digital signal processing. *Proceedings of IEEE International Conference on Acoustics, Speech, and Signal Processing*, **5**, 2594–7.

ERIKSSON, L.J., ALLIE, M.C., BREMIGAN, C.D., and GILBERT, J.A. (1988*b*). Active noise control and specifications for fan noise problems. *Proceedings of Noise-Con 88*, pp. 273–8.

FEDORYUK, M.V. (1975*a*). An active noise-suppression method. *Soviet Physics Acoustics*, English translation, **20**, (5), 495–6.

FEDORYUK, M.V. (1975*b*). The suppression of sound in acoustic waveguides. *Soviet Physics Acoustics*, English translation, **21**, (2), 174–6.

FEDORYUK, M.V. (1976). The nonsteady active noise-suppression problem. *Soviet Physics Acoustics*, English translation, **22**, (3), 244–6.

FEDORYUK, M.V. (1979). Active sound suppression arrays of monopoles. *Soviet Physics Acoustics*, English translation, **25**, (1), 61–4.

FLOCKTON, S.J. (1987). Cancellation of acoustic noise in a pipe using digital adaptive filters. *Proceedings of IEEE International Conference on Acoustics, Speech, and Signal Processing*, Dallas, **1**, 165–8.

FORD, R.D. (1984). Rower requirements for active noise control in ducts. *Journal of Sound and Vibration*, **92**, (3), 411–8.

FRY, A. and FRYER, P. (1985). Loudspeakers for sound cancellation. *Proceedings of the Institute of Acoustics*, **7**, Part 3, 261–5.

F-WILLIAMS, J.E. (1982). The silent noise of a gas turbine. *Elektor*, **4**, 14–16.

F-WILLIAMS, J.E. (1984). Anti-sound. *Proceedings of the Royal Society of London*, Series A, **395**, 63–88.

F-WILLIAMS, J.E. (1985). Anti-noise. *Proceedings of National Conference on Noise Control Engineering*, Ohio, pp. 3–6.

F-WILLIAMS, J.E., ROEBUCK, I., and ROSS, C.F. (1985). Anti-phase noise reduction. *Physical Technology*, **16**, 19–24, 31.

GAN, W.S. (1984). Application of active noise absorption to infrasonics. *Proceedings of Inter-noise 84: International Conference on Noise Control Engineering*, Honolulu, **I**, 453–6.

GAUDEFROY, A. (1981*a*). Determination of the configurations of active sound attenuation in a duct by means of the theory of distributions. *Acoustics Letters*, **4**, (7), 136–44.

GAUDEFROY, A. (1981*b*). Determination of the configurations of active attenuation sources in a duct by means of an optimal method. *Acoustics Letters*, **4**, (7), 145–9.

GAVRILOV, A.M., LYUBASHEVSKII, G.S., and CHONI, YU.I. (1977). Effectiveness rating of criteria for the placement of the compensating sources in active suppression systems. *Soviet Physics Acoustics*, English translation, **23**, (3), 269–70.

GLENDINNING, A., NELSON, P.A., and ELLIOTT, S.J. (1985). A study of sound generation using electro-pneumatic sources. *Proceedings of the Institute of Acoustics*, **7**, Part 3, 283–90.

GUICKING, D. and ROLLWAGE, M. (1984). Active systems in room acoustics – solved and unsolved problems. *Proceedings of Inter-noise 84: International Conference on Noise Control Engineering*, Honolulu, **I**, 457–62.

GUICKING, D., KARCHER, K., and ROLLWAGE, M. (1983). Active control the acoustic reflection coefficient at low frequencies. *Proceedings of Inter-noise 83: International Conference on Noise Control Engineering*, Edinburgh, **I**, 419–22.

HARRISON, W.A., LIM, J.S., and SINGER, E. (1986). A new application of adaptive noise cancellation. *IEEE Transactions on Acoustics, Speech, and Signal Processing*, **34**, (1), 21–7.

HAWLEY, M.E. and SIMSHAUSER, E.D. (1961). *Noise reducing system.* US Patent No. 2 972 018.

HECKL, M. (1984). Active control of the noise of a rijke tube. *Proceedings of the Institute of Acoustics*, **6**, Part 1, 219–21.

HESSELMAN, N. (1978). Investigation of noise reduction on a 100 KVA transformer tank by means of active methods. *Applied Acoustics*, **11**, (1), 27–34.

HONG, W.K.W. and EGHTESADI, KH. (1983). Applications of active attenuator in controlling low-frequency noise. *Journal of Low Frequency Noise and Vibration*, **2**, (1), 29–36.

HONG, W.K.W., EGHTESADI, KH., and LEVENTHAL, H.G. (1982). The tandem tight-coupled active noise attenuator. *Acoustics Letters*, **6**, (2), 19–24.

HONG, W.K.W., EGHTESADI, KH., and LEVENTHAL, H.G. (1983). The tight-coupled monopole (TCM) and tandem (TCT) attenuators in a duct. *Proceedings of Inter-noise 83: International Conference on Noise Control Engineering*, **I**, 439–42.

HONG, W.K.W., EGHTESADI, KH., and LEVENTHAL, H.G. (1985). Energy conservation by active noise attenuation in ducts. *Proceedings of the Institute of Acoustics*, **7**, Part 2, 103–9.

HONG, W.K.W., EGHTESADI, KH., and LEVENTHAL, H.G. (1987). The tight-coupled monopole (TCM) and tight-coupled tandem (TCT) attenuators: theoretical aspects and experimental attenuation in an air duct. *Journal of the Acoustical Society of America*, **81**, (2), 376–88.

IVANOV, YA.I. (1985). Active suppression of sound fields in an air duct. *Soviet Physics Acoustics*, English translation, **31**, (3), 237–8.

JESSEL, M. (1979). Some evidence for a general theory of active sound absorption. *Proceedings of Inter-noise 79: International Conference on Noise Control Engineering*, Warsaw, **I**, 169–74.

JESSEL, M.J.M. (1981). Secondary sources and their energy transfer. *Acoustics Letters*, **4**, (9), 174–9.

JESSEL, M.J.M. (1983). Active noise reduction as an experimental application of general system theory. *Proceedings of Inter-noise 83: International Conference on Noise Control Engineering*, Edinburgh, **I**, 411–14.

JESSEL, M. (1985). Active noise and vibration control (ANVC): current trends, permanent aims and future possibilities. *Archives of Acoustics*, **10**, (4), 345–56.

JESSEL, M. and ANGEVINE, O.L. (1980). Active acoustic attenuation of a complex noise source. *Proceedings of Inter-noise 80: International Conference on Noise Control Engineering*, Florida, **II**, 689–94.

JESSEL, M. and MANGIANTE, G.A. (1972). Active sound absorbers in an air duct. *Journal of Sound and Vibration*, **23**, (3), 383–90.

JESSEL, M. and YAMADA, S. (1987). Active noise control. *Journal of the Acoustical Society of Japan (E)*, **8**, (4), 151–4.

KAZAKIA, J.Y. (1986). A study of active attenuation of broadband noise. *Journal of Sound and Vibration*, **110**, (3), 495–509.

KEMPTON, A.J. (1976). The ambiguity of acoustic sources – a possibility of active control. *Journal of Sound and Vibration*, **48**, (4), 475–83.

KIDO, K. (1975). Reduction of noise by use of additional sound sources. *Proceedings of Inter-noise 75: International Conference on Noise Control Engineering*, Sendai, Japan, pp. 647–50.

KIDO, K., ABE, M., and MORIKAWA, S. (1984). Suppression of acoustic noise by computer generated sound. *Proceedings of Inter-noise 84: International Conference on Noise Control Engineering*, Honolulu, **I**, 463–8.

KIM, C., ABE, M., and KIDO, K. (1983). Cancellation of signal picked up in a room by estimated signal. *Proceedings of Inter-noise 83: International Conference on Noise Control Engineering*, Edinburgh, **I**, 423–6.

KIM, C.D., MORIKAWA, S., ABE, M., and KIDO, K. (1984). Active suppression of noise in microphone line using vibration of noise source. *Proceedings of Inter-noise 84: International Conference on Noise Control Engineering*, Honolulu, **I**, 469–72.

KLIMOV, S.P., MAZANIKOV, A.A., MERKULOV, V.N., and TYUTEKIN, V.V. (1980). Ultimate bandwidth of an active sound-suppression system in a single-mode waveguide. *Soviet Physics Acoustics*, English translation, **26**, (1), 82.

KOERS, P. (1986). Adaptive, active reduction of periodic noise using a frequency domain LMS algorithm. *Proceedings of Inter-noise 86: International Conference on Noise Control Engineering*, Cambridge, Mass., **I**, 595–600.

KONYAEV, S.I., LEBEDEV, V.I., and FEDORYUK, M.V. (1977). Discrete approximation of a spherical Huygens surface. *Soviet Physics Acoustics*, English translation, **23**, (4), 373–4.

KONYAEV, S.I., LEBEDEV, V.I., and FEDORYUK, M.V. (1979). Factorization of a sound field by means of two concentric spherical surfaces. *Soviet Physics Acoustics*, English translation, **25**, (5), 413–15.

KOROTAEV, E.V. and MAZANIKOV, A.A. (1985). Active sound suppression by a finite planar array. *Soviet Physics Acoustics*, English translation, **31**, (4), 323–4.

KOSAKA, T., ISHIZAWA, M., MIYANO, H., and YAMADA, S. (1984). Active noise reduction of low frequency noise in a pipe. *Proceedings of Inter-noise 84: International Conference on Noise Control Engineering*, Honolulu, **I**, 473–6.

KROLL, M.W. (1987). *Lung sound cancellation method and apparatus.* US Patent No. 4 672 977.

LaFontaine, R.F. and Shepherd, I.C. (1983). An experimental study of broadband active attenuator for cancellation of random noise in ducts. *Journal of Sound and Vibration*, **91**, (3), 351–62.

LaFontaine, R.F. and Shepherd, I.C. (1985). The influence of waveguide reflections and system configuration on the performance of an active noise attenuator. *Journal of Sound and Vibration*, **100**, (4), 569–79.

Lamancusa, J.S. (1987). An actively tuned, passive muffler system for engine silencing. *Proceedings of National Conference on Noise Control Engineering*, Pennsylvania, pp. 313–18.

Leitch, R.R. and Tokhi, M.O. (1986). The implementation of active noise control systems using digital signal processing techniques. *Proceedings of the Institute of Acoustics*, **8**, Part 1, 149–57.

Leitch, R.R. and Tokhi, M.O. (1987). Active noise control systems. *IEE Proceedings*, Part A, **134**, (6), 525–46.

Leitch, R.R., Alvarez, A., and Yang, S.J. (1983). Spectral conditions for the active cancellation of broadband noise. *Proceedings of Inter-noise 83: International Conference on Noise Control Engineering*, Edinburgh, **I**, 443–6.

Lester, H.C. and Fuller, C.R. (1987). Mechanisms of active control for noise inside a vibrating cylinder. *Proceedings of National Conference on Noise Control Engineering*, Pennsylvania, pp. 371–6.

Leventhal, H.G. (1976). Developments in active attenuators. *Proceedings of Noise Control Conference*, Warsaw, pp. 33–42.

Leventhal, H.G. (1979). Active attenuators for noise. *Proceedings of the Institute of Acoustics*, Autumn, 193–6.

Leventhal, H.G. (1980). Active attenuators – an historical review and some recent developments. *Proceedings of Inter-noise 80: International Conference on Noise Control Engineering*, Florida, **II**, 679–82.

Leventhal, H.G. (1981). Active attenuators: a historical review and some recent developments. *Proceedings of the Institute of Acoustics*, February, 1–4.

Leventhal, H.G. and Eghtesadi, Kh. (1979). Active attenuation of noise: monopole and dipole systems. *Proceedings of Inter-noise 79: International Conference on Noise Control Engineering*, Warsaw, **I**, 175–80.

Levine, R.C. (1980). False paradoxes of superposition in electric and acoustic waves. *American Journal of Physics*, **48**, (1), 28–31.

Lueg, P. (1936). *Process of silencing sound oscillations.* US Patent No. 2 043 416.

Martin, V. and Roure, A. (1983). Control of sources for active sound propagation in a duct. *Proceedings of Inter-noise 83: International Conference on Noise Control Engineering*, Edinburgh, **I**, 431–4.

Mazanikov, A.A. and Tyutekin, V.V. (1975). Experimental study of an active system for the suppression of sound fields. *Soviet Physics Acoustics*, English translation, **20**, (5), 493.

Mazanikov, A.A. and Tyutekin, V.V. (1976). Autonomous active systems for the suppression of sound fields in single-mode waveguides. *Soviet Physics Acoustics*, English translation, **22**, (5), 409–12.

Mazanikov, A.A. and Tyutekin, V.V. (1978). Efficiency and stability of autonomous active-suppression systems. *Soviet Physics Acoustics*, English translation, **24**, (5), 447–9.

MAZANIKOV, A.A., TYUTEKIN, V.V., and UKOLOV, A.T. (1977*a*). An active system for the suppression of sound fields in a multimode waveguide. *Soviet Physics Acoustics*, English translation, **23**, (3), 276–7.

MAZANIKOV, A.A., UKOLOV, A.T., and FEDORYUK, M.V. (1977*b*). Active suppression of limited-frequency sound in waveguides. *Soviet Physics Acoustics*, English translation, **23**, (6), 519–21.

MAZANIKOV, A.A., TYUTEKIN, V.V., and FEDORYUK, M.V. (1980). Active suppression of sound fields by the method of spatial harmonics. *Soviet Physics Acoustics*, English translation, **26**, (5), 428–30.

MAZZANTI, S. and PIRAUX, J. (1983). An experiment of active noise attenuation in three-dimensional space. *Proceedings of Inter-noise 83: International Conference on Noise Control Engineering*, Edinburgh, **I**, 427–30.

MERKULOV, V.N. and TYUTEKIN, V.V. (1980). Active suppression of flexural modes in a fluid-filled cylindrical shell. *Soviet Physics Acoustics*, English translation, **26**, (4), 354–5.

MICHALKE, A. (1987). A note on non-compact acoustic monopoles and dipoles. *Journal of Sound and Vibration*, **117**, (3), 519–27.

MULLER, G.S. and PAUW, C.K. (1986). Acoustic noise cancellation. *Proceedings of IEEE International Conference on Acoustics, Speech, and Signal Processing*, Tokyo, **2**, 913–16.

MUNJAL, M.L. and ERIKSSON, L.J. (1988). An analytical one-dimensional, standing-wave model of a linear active noise control system in a duct. *Journal of the Acoustical Society of America*, **84**, (3), 1086–93.

NADIM, M. and SMITH, R.A. (1983). Synchronous adaptive cancellation in vehicle cabins. *Proceedings of Inter-noise 83: International Conference on Noise Control Engineering*, Edinburgh, **I**, 461–4.

NARAYANA DUTT, D. and RAMAKRISHNA, B.A. (1978). An experimental study of controlling the sound distribution in a tube by using auxiliary driving sources. *Journal of Sound and Vibration*, **60**, (2), 267–71.

NELSON, P.A., CURTIS, A.R.D., and ELLIOTT, S.J. (1985). Quadratic optimisation problems in the active control of free and enclosed sound fields. *Proceedings of the Institute of Acoustics*, **7**, Part 2, 45–53.

NELSON, P.A., CURTIS, A.R.D., and ELLIOTT, S.J. (1986). On the active absorption of sound. *Proceedings of Inter-noise 86: International Conference on Noise Control Engineering*, Cambridge, Mass., **I**, 601–6.

NELSON, P.A., CURTIS, A.R.D., ELLIOTT, S.J., and BULLMORE, A.J. (1987*a*). The minimum power output of free field point sources and the active control of sound. *Journal of Sound and Vibration*, **116**, (3), 397–414.

NELSON, P.A., CURTIS, A.R.D., ELLIOTT, S.J., and BULLMORE, A.J. (1987*b*). The active minimization of harmonic enclosed sound fields, Part I: Theory. *Journal of Sound and Vibration*, **117**, (1), 1–13.

NELSON, P.A., HAMMOND, J.K., and ELLIOTT, S.J. (1987*c*). Causal constraints on the active control of sound. *Proceedings of IEEE International Conference on Acoustics, Speech, and Signal Processing*, Dallas, **I**, 145–8.

OKDA, J., USAGAWA, T., and EBATA, M. (1987). Analysis of active control of noise in duct by means of impedance. *Journal of the Acoustical Society of Japan (E)*, **8**, (2), 63–72.

OLSON, H.F. and MAY, E.G. (1953). Electronic sound absorbers. *Journal of the Acoustical Society of America*, **25**, (6), 1130–6.

OLSON, H.F. (1956). Electronic control of noise, vibration and reverberation. *Journal of the Acoustical Society of America*, **28**, (5), 966–72.

OLSON, H.F. (1961). *Electronic sound absorber*. US Patent No. 2 983 790.

OSWALD, L.J. (1984). Reduction of diesel engine noise inside passenger compartments using active, adaptive noise control. *Proceedings of Inter-noise 84: International Conference on Noise Control Engineering*, Honolulu, **I**, 483–8.

PARRAMORE, T.S. (1974). *Improvements in or relating to dynamic silencing systems*. UK Patent No. 1 357 330.

PIRAUX, J. and MAZZANTI, S. (1985). Broadband active noise attenuation in three-dimensional space. *Proceedings of Inter-noise 85: International Conference on Noise Control Engineering*, Munich, **I**, 485–8.

PIRAUX, J. and NAYROLES, B. (1980). A theoretical model for active noise attenuation in three-dimensional space. *Proceedings of Inter-noise 80: International Conference on Noise Control Engineering*, Florida, **II**, 703–6.

POOLE, J. and LEVENTHAL, G.H. (1976). An experimental study of Swinbanks' method of active attenuation of sound in ducts. *Journal of Sound and Vibration*, **49**, (2), 257–66.

POOLE, J.H.B. and LEVENTHAL, H.G. (1978). Active attenuators of noise in ducts. *Journal of Sound and Vibration*, **57**, (2), 308–9.

POOLE, L.A. and WARNAKA, G.E. (1987). Improved active noise control microphone. *Proceedings of National Conference on Noise Control Engineering*, Pennsylvania, pp. 377–82.

POOLE, L.A., WARNAKA, G.E., and CUTTER, R.C. (1984). The implementation of digital filters using modified Widrow–Hoff algorithm for the adaptive cancellation of acoustic noise. *Proceedings of IEEE International Conference on Acoustics, Speech, and Signal Processing*, California, **II**, 21.7.1–3.

POWELL, G.A., DARLINGTON, P., and WHEELER, P.D. (1987). Practical adaptive noise reduction in the aircraft cockpit environment. *Proceedings of IEEE International Conference on Acoustics, Speech, and Signal Processing*, Dallas, **1**, 173–6.

RICHARZ, W.G. (1984). The wavy wall concept and active noise control in ducts. *Proceedings of Inter-noise 84: International Conference on Noise Control Engineering*, Honolulu, **I**, 489–92.

RICHARZ, W.G. (1986). Phased array active noise controller for ducts. *Journal of the Acoustical Society of America*, Supplement 1, **80**, s11.

ROBERTS, J.P. and VUILLERMOZ, M.L. (1987). *Active control of acoustic instability in combustion chambers*. US Patent No. 4 644 783.

RODRIGUEZ, J.J., LIM, J.S., and SINGER, E. (1987). Adaptive noise reduction in aircraft communication systems. *Proceedings of IEEE International Conference on Acoustics, Speech, and Signal Processing*, Dallas, **1**, 169–72.

ROEBUCK, I. (1981). Stability considerations and performance limits in multiple source cancellation. *Proceedings of the Institute of Acoustics*, February, 17–20.

ROSS, C.F. (1978). Experiments on the active control of transformer noise. *Journal of Sound and Vibration*, **61**, (4), 473–80.

ROSS, C.F. (1981*a*). Design of a broadband active sound control system. *Proceedings of the Institute of Acoustics*, February, 21–4.

ROSS, C.F. (1981*b*). A demonstration of active control of broadband sound. *Journal of Sound and Vibration*, **74**, (3), 411–18.

ROSS, C.F. (1981*c*). *Process of testing for a sound control system*. UK Patent No. 2 069 280.

ROSS, C.F. (1982*a*). Application of digital signal filtering to active control of sound. *Acustica*, **51**, (2), 135–40.

ROSS, C.F. (1982*b*). An algorithm for designing a broadband active sound control system. *Journal of Sound and Vibration*, **80**, (3), 373–80.

ROSS, C.F. (1982*c*). An adaptive digital filter for broadband active sound control. *Journal of Sound and Vibration*, **80**, (3), 381–8.

ROSS, C.F. (1982*d*). Active noise control. *Proceedings of the Institute of Acoustics*, May, 35–7.

ROSS, C.F. (1982*e*). Quietening newspaper printing presses by active control. *Proceedings of the Institute of Acoustics*, November, F1.1–F1.4.

ROSS, C.F. (1986). Novel applications of active control techniques. *Proceedings of the Institute of Acoustics*, **8**, Part 1, 127–33.

ROURE, A. (1985). Self-adaptive broadband active sound control system. *Journal of Sound and Vibration*, **101**, (3), 429–41.

ROURE, A. and NAYROLES, B. (1984). Autoadaptive broadband active absorption in ducts by the means of transversal filtering. *Proceedings of Inter-noise 84: International Conference on Noise Control Engineering*, Honolulu, **I**, 493–6.

SHEPHERD, I.C., LAFONTAINE, R.F., and CABELLI, A. (1984). Active attenuation in turbulent flow ducts. *Proceedings of Inter-noise 84: International Conference on Noise Control Engineering*, Honolulu, **I**, 497–502.

SHEPHERD, I.C., CABELLI, A., and LAFONTAINE, R.F. (1986). Characteristics of loudspeakers operating in an active noise attenuator. *Journal of Sound and Vibration*, **110**, (3), 471–81.

SHORT, W.R. (1980). Global low frequency active noise attenuation. *Proceedings of Inter-noise 80: International Conference on Noise Control Engineering*, Florida, **II**, 695–8.

SILCOX, R.J. and ELLIOTT, S.J. (1985). Applicability of superposition and source impedance models of active noise control systems. *Proceedings of Inter-noise 85: International Conference on Noise Control Engineering*, Munich, **I**, 587–90.

SIMSHAUSER, E.D. and HAWLEY, M.E. (1955). The noise cancelling headset – an active ear defender. *Journal of the Acoustical Society of America*, **27**, (1), 207.

SIMSHAUSER, E.D., MEEKER, W.F., and BALAKRISHNAN, A.V. (1956). Noise cancelling headset – an active ear defender. *Journal of the Acoustical Society of America*, **28**, (4), 773.

SJOSTEN, P. (1985). Experiences of an adaptive control system for active noise control. *Proceedings of Inter-noise 85: International Conference on Noise Control Engineering*, Munich, **I**, 595–8.

SJOSTEN, P., ERIKSSON, P., and CLAESSON, I. (1986). Computer simulation in active noise control. *Proceedings of the Institute of Acoustics*, **8**, Part 1, 187–90.

SMITH, R.A. and CHAPLIN, G.B.B. (1983). A comparison of some Essex algorithms for major industrial applications. *Proceedings of Inter-noise 83: International Conference on Noise Control Engineering*, Edinburgh, **I**, 407–10.

SWANSON, D.C. (1987). A simulation of computer fan noise cancellation using adaptive armax control. *Proceedings of National Conference on Noise Control Engineering*, Pennsylvania, pp. 383–8.

SWINBANKS, M.A. (1973). Active control of sound propagation in long ducts. *Journal of Sound and Vibration*, **27**, (3), 411–36.

SWINBANKS, M.A. (1976). *Active control of sound waves.* UK Patent No. 1 456 018.

SWINBANKS, M.A. (1977). *Active control of sound waves.* US Patent No. 4 044 203.

SWINBANKS, M.A. (1979). *Active control of sound waves.* UK Patent No. 1 548 362.

SWINBANKS, M.A. (1981). *Signal processing systems.* UK Patent No. 2 054 999.

SWINBANKS, M.A. (1982). The active control of low frequency sound in a gas turbine compressor installation. *Proceedings of Inter-noise 82: International Conference on Noise Control Engineering*, San Francisco, **II**, 423–6.

SWINBANKS, M.A. (1986). An overall perspective of active noise control: the balance between theoretical and practical research. *Proceedings of the Institute of Acoustics*, **8**, Part 1, 117–26.

SWINBANKS, M.A. (1987). *Methods and apparatus for reducing noise by cancellation.* US Patent No. 4 637 048.

TAKAHASHI, M., KURIBAYASHI, T., ASAMI, K., HAMADA, H., ENOKIDA, T., and MIURA, T. (1986). Electric sound cancellation in air-conditioning duct system. *Proceedings of Inter-noise 86: International Conference on Noise Control Engineering*, Cambridge, Mass., **I**, 607–10.

TAKAHASHI, M., KURIBAYASHI, T., ASAMI, K., ENOKIDA, T., HAMADA, H., and MIURA, T. (1987). Broadband active sound control system for air-conditioning duct noise. *Journal of the Acoustical Society of Japan (E)*, **8**, (6), 263–9.

TARTAKOVSKII, B.D. (1975). Active methods for the compensation of vibroacoustic fields. *Soviet Physics Acoustics*, English translation, **20**, (5), 494–5.

TAYLOR, B.A. (1984). Contranoise – active noise control. *Noise and Vibration Control Worldwide*, **15**, (9), 256–9.

TICHY, J. and WARNAKA, G.E. (1983). Effects of evanescent waves on the active attenuation of sound in ducts. *Proceedings of Inter-noise 83: International Conference on Noise Control Engineering*, Edinburgh, **I**, 435–8.

TICHY, J., POOLE, L.A., and WARNAKA, G.E. (1987). Requirements for active and passive noise control in hearing protectors. *Proceedings of Noise-Con. 87: National Conference on Noise Control Engineering*, Pennsylvania, pp. 389–92.

TOHYAMA, M. and SUZUKI, A. (1987). Active power minimization of a sound source in a closed space. *Journal of Sound and Vibration*, **119**, (3), 562–4.

TOKHI, M.O. and LEITCH, R.R. (1988). Practical limitations in the controller design for active noise control systems in three dimensions. *Proceedings of Inter-noise 88: International Conference on Noise Control Engineering*, Avignon, **2**, 1037–40.

TOKHI, M.O. and LEITCH, R.R. (1989). Self-tuning active noise control. In *Adaptive filters*, IEE Colloquium, Digest No. 1989/16, 9/1–4.

TRINDER, M.C. and JONES, O. (1987). Active noise control at the ear. *Proceedings of National Conference on Noise Control Engineering*, Pennsylvania, pp. 393–8.

TRINDER, M.C.J. and NELSON, P.A. (1983*a*). The acoustical virtual earth and its application to ducts with reflecting terminations. *Proceedings of Inter-noise 83: International Conference on Noise Control Engineering*, Edinburgh, **I**, 447–50.

TRINDER, M.C.J. and NELSON, P.A. (1983*b*). Active noise control in finite length ducts. *Journal of Sound and Vibration*, **89**, (1), 95–106.

TRINDER, M.C.J., CHAPLIN, G.B.B., and NELSON, P.A. (1986). Active control of commercial vehicle exhaust noise. *Proceedings of Inter-noise 86: International Conference on Noise Control Engineering*, Cambridge, Mass., **I**, 611–16.

TWINEY, R.C., HOLDEN, A.J., and SALLOWAY, A.J. (1985). Some transducer design considerations for earphone active noise reduction systems. *Proceedings of the Institute of Acoustics*, **7**, Part 2, 95–102.

URUSOVSKII, I.A. (1977*a*). Active noise suppression in a waveguide. *Soviet Physics Acoustics*, English translation, **23**, (2), 170–4.

URUSOVSKII, I.A. (1977*b*). Self-excitation of an active noise-suppression system in a waveguide. *Soviet Physics Acoustics*, English translation, **23**, (3), 243–6.

URUSOVSKII, I.A. (1980). Active noise suppression in a waveguide with monopole radiators and dipole receivers. *Soviet Physics Acoustics*, English translation, **26**, (2), 153–7.

UROSOVSKII, I.A. (1981). Active sound suppression by monopole distributed over a single surface. *Soviet Physics Acoustics*, English translation, **27**, (4), 325–9.

URUSOVSKII, I.A. (1986). Active synthesis of a scattered sound field. *Soviet Physics Acoustics*, English translation, **32**, (4), 354–5.

VANDENHOUT, J. and SAS, P. (1986). Active noise cancellation: a feasibility study. *Proceedings of the Institute of Acoustics*, **8**, Part 1, 165–72.

VOGT, M. (1976). General conditions of phase cancellation in an acoustic field. *Archives of Acoustics*, **1**, (2), 109–25.

VYALYSHEV, A.I., GAVRILOV, A.M., LYUBASHEVSKII, G.S., TARTAKOVSKII, B.D., and CHONI, YU.I. (1977). Synthesis of systems for the suppression of vibration and sound fields. *Soviet Physics Acoustics*, English translation, **23**, (2), 136–9.

VYALYSHEV, A.I., DUBININ, A.I., and TARTAKOVSKII, B.D. (1986). Active acoustic reduction of a plate. *Soviet Physics Acoustics*, English translation, **32**, (2), 96–8.

VYALYSHEV, A.I., DUBININ, A.I., and TARTAKOVSKII, B.D. (1987). Active noise reduction of a cylindrical shell. *Soviet Physics Acoustics*, English translation, **33**, (1), 63–5.

WARNAKA, G.E. (1981). Improvements in adaptive active attenuators. *Proceedings of Inter-noise 81: International Conference on Noise Control Engineering*, **1**, 307–10.

WARNAKA, G.E. (1982*a*). Active attenuation of noise: the state of the art. *Noise Control Engineering*, **18**, (3), 100–10.

WARNAKA, G.E. (1982*b*). *Acoustic attenuators with active sound cancelling*. UK Patent No. 2 088 951.

WARNAKA, G.E. (1987). Applications for active noise control. *Proceedings of National Conference on Noise Control Engineering*, Pennsylvania, pp. 399–404.

WARNAKA, G.E. and TICHY, J. (1980). Acoustic mixing in active attenuators. *Proceedings of Inter-noise 80: International Conference on Noise Control Engineering*, Florida, **II**, 683–8.

WARNAKA, G.E., POOLE, L.A., and TICHY, J. (1982). Transient response in active systems. *Proceedings of Inter-noise 82: International Conference on Noise Control Engineering*, San Francisco, **II**, 427–30.

WARNAKA, G.E., ZALAS, J.M., TICHY, J., and POOLE, L.A. (1983). Active control of noise in interior spaces. *Proceedings of Inter-noise 83: International Conference on Noise Control Engineering*, Edinburgh, **I**, 415–18.

WHEELER, P.D. and HALLIDAY, S.G. (1981). An active noise reduction system for aircrew helmets – flight trials in strike aircraft. *Proceedings of the Institute of Acoustics*, February, 5–8.

WHEELER, P.D. and PARRAMORE, T.S. (1981). Active silencing of a transportable generator set. *Proceedings of the Institute of Acoustics*, February, 25–8.

WHEELER, P.D., RAWLINSON, R.D., PELC, S.F., and DOREY, A.P. (1978). The development and testing of an active noise reduction system for use in ear defenders. *Proceedings of Inter-noise 78: International Conference on Noise Control Engineering*, San Francisco, pp. 977–82.

WURM, R.J. and BERGSON, A.A. (1974). *Noise cancellation*. US Patent No. 3 826 870.

YAMADA, S., JESSEL, M., and ANGELINI, TH. (1979). Active sound absorption by a single monopole source for the reduction of very low frequency sound in a duct. *Acoustics Letters*, **3** (2), 56–9.

ZAVADSKAYA, M.P. and URUSOVSKII, I.A. (1976). Influence of random errors on the degree of compensation of sound fields in an active noise-suppression problem. *Soviet Physics Acoustics*, English translation, **22**, (2), 127–30.

ZAVADSKAYA, M.P., POPOV, A.V., and EGEL'SKII, B.L. (1975*a*). Approximation of wave potentials in the active suppression of sound fields by the Malyuzhinets method. *Soviet Physics Acoustics*, English translation, **21**, (5), 451–4.

ZAVADSKAYA, M.P., POPOV, A.V., and EGEL'SKII, B.L. (1975*b*). An approximate solution of the problem of active suppression of sound fields by the Malyuzhinets method. *Soviet Physics Acoustics*, English translation, **21**, (6), 541–4.

ZAVADSKAYA, M.P., POPOV, A.V., and EGEL'SKII, B.L. (1977). Approximation and stability aspects of active suppression systems with a finite number of couplings. *Soviet Physics Acoustics*, English translation, **23**, (3), 272–3.

3

THE PROCESS OF FIELD CANCELLATION

3.1 Introduction

Active noise cancellation is based upon the intentional superposition of acoustic waves to result in destructive interference. This is realized by artificially generating sound waves that interfere with the noise, thereby using the destructive interference of the component waves to reduce the level of noise. There are a number of factors that influence the extent of noise cancellation, among these are: the artificially generated (secondary) source as related to the unwanted (primary) source, physical separation between the sources, and properties of the propagation medium.

To achieve cancellation at a given point in the medium it is essential that the primary and secondary waves be coherent and in anti-phase at the point. The former implies that there must be a constant phase difference between component waves of the same frequency. The latter means that the two waves should interfere destructively with one another. If the two waves are uncorrelated then stable cancellation is not possible. However, if they are coherent but not in anti-phase then the interference may be constructive leading to a reinforcement of the primary wave and an overall increase in the noise level.

In propagating through the medium the waves are affected in amplitude as well as in phase due to the properties of the medium. This determines whether the interference at a specified point in the medium is destructive or constructive. Here, assuming a linear propagation medium, the separation between the sources plays a major role.

The above issues are discussed and verified in this chapter through a parametric analysis of the process of field cancellation in a three-dimensional propagation medium. The analysis is based on two compact (point) sources operating in a non-dispersive (linear) propagation medium under stationary or slowly varying conditions. A general analysis of the process of field cancellation, from a time-domain and a frequency-domain perspective, leading to the basic conditions of cancellation has been considered previously (Leitch and Tokhi 1987; Vogt 1976). Others have given a description of cancellation in relation to specific applications (Leventhal 1976).

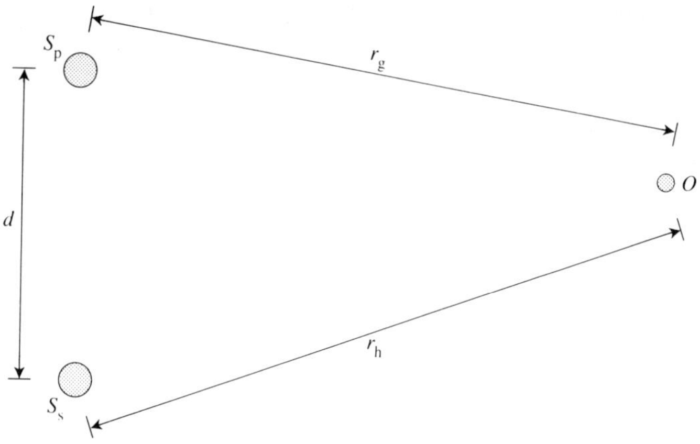

Fig. 3.1. Basic schematic diagram of active noise control

The analysis in this chapter is presented from the perspective of the design of ANC systems for broadband noise. The general conditions necessary for noise cancellation are developed. For a single tone of the component waves a quantitative measure of the degree of cancellation, in terms of amplitude and phase parameters of the waves, is given. In the case of broadband noise this is extended to provide relationships for the degree of cancellation as functions of frequency. Geometry-related conditions involving the separation between the sources and location of a point in the medium at which cancellation is required are obtained. These lead to a three-dimensional description of cancellation in the medium. This is useful for determining suitable geometrical arrangements of system components for a given degree of cancellation.

3.2 The field cancellation factor

In this section a quantitative measure of the degree of cancellation is developed. This is given in terms of the amplitude and phase parameters of the component waves.

Let a (primary) source S_p emit a wave $p(t)$ as a function of time t into a non-dispersive (linear) propagation medium. This, in propagating through the medium, results in a field $p(r, t)$ at a distance r from S_p. Thus, at a fixed point O in the medium, at a distance r_g from the primary source, the wave $p(t)$ will give rise to a field $p(r_g, t)$, or simply $p_o(t)$, that is a function of t only. As depicted in Fig. 3.1, let a (secondary) source S_s be placed at a distance d from S_p and emit a wave $s(t)$. This, in propagating through a distance r_h from S_s, will give rise to $s_o(t)$ at point O. The component waves from S_p and S_s will thus result in a (combined or observed) signal $o(t)$ at point O. Let the power spectral densities of the waves be denoted as follows (where ω represents the radian frequency):

$G_{pp}(\omega)$ = auto power spectral density of the primary wave $p(t)$,
$G_{ss}(\omega)$ = auto power spectral density of the secondary wave $s(t)$,
$G_{ppo}(\omega)$ = auto power spectral density of the primary wave $p_o(t)$,
$G_{sso}(\omega)$ = auto power spectral density of the secondary wave $s_o(t)$, and
$G_{cco}(\omega)$ = auto power spectral density of the observed wave $o(t)$.

Cancellation at the observation point O (of the primary wave) occurs when the power spectral density of the combined wave is less than that of the primary wave acting alone:

$$G_{cco}(\omega) < G_{ppo}(\omega). \tag{3.1}$$

If the secondary wave is desired to be cancelled then the following relation should hold

$$G_{cco}(\omega) < G_{sso}(\omega).$$

For a quantitative description of the phenomenon of cancellation the field cancellation factor K is defined as the ratio of the cancelled spectrum, $G_{ppo}(\omega) - G_{cco}(\omega)$, to the primary wave spectrum, $G_{ppo}(\omega)$, which existed at the given point O prior to the superposition of the secondary field, $s_o(t)$:

$$K \triangleq \frac{G_{ppo}(\omega) - G_{cco}(\omega)}{G_{ppo}(\omega)} \tag{3.2}$$

or

$$K = 1 - \frac{G_{cco}(\omega)}{G_{ppo}(\omega)}. \tag{3.3}$$

Note that here the primary field is assumed to be cancelled.

A frequency-domain description of Fig. 3.1 is given in Fig. 3.2 where $P(j\omega)$,

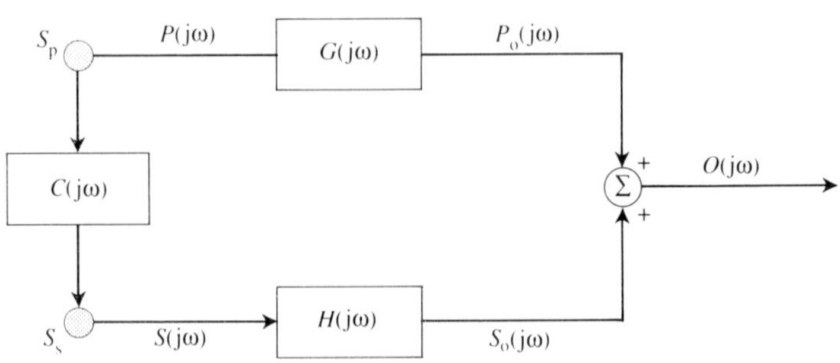

Fig. 3.2. Block diagram of the basic ANC structure

$P_o(j\omega)$, $S(j\omega)$, $S_o(j\omega)$, and $O(j\omega)$ represent the corresponding signals $p(t)$, $p_o(t)$. $s(t)$, $s_o(t)$, and $o(t)$, respectively. $C(j\omega)$ represents a frequency-dependent transfer function, with amplitude $H_c(\omega)$ and phase $\theta_c(\omega)$, through which the secondary signal $s(t)$ is derived from the primary signal $p(t)$:

$$C(j\omega) = H_c(\omega)e^{j\theta_c(\omega)}. \tag{3.4}$$

$G(j\omega)$ and $H(j\omega)$ are the transfer characteristics of the acoustic paths from S_p and S_s, respectively, through the distances r_g and r_h to the observation point O. Assuming a non-dispersive propagation medium, these are given by

$$\begin{aligned} G(j\omega) &= \frac{A}{r_g} e^{-j2\pi r_g/\lambda} = \frac{A}{r_g} e^{-j\omega r_g/c} \\ H(j\omega) &= \frac{A}{r_h} e^{-j2\pi r_h/\lambda} = \frac{A}{r_h} e^{-j\omega r_h/c} \end{aligned} \tag{3.5}$$

where λ is the wavelength, c the velocity of sound in the medium, and A is a constant. The ratio $2\pi/\lambda$ is known as the wave number, k:

$$k = \frac{2\pi}{\lambda} = \frac{\omega}{c}. \tag{3.6}$$

The ratios r_g/c and r_h/c in equation (3.5) are, respectively, the times t_g and t_h at which the primary and secondary waves reach the observation point O in propagating, respectively, through the distances r_g and r_h:

$$\begin{aligned} t_g &= \frac{r_g}{c} \\ t_h &= \frac{r_h}{c}. \end{aligned} \tag{3.7}$$

It follows from Fig. 3.2 that

$$\begin{aligned} P_o(j\omega) &= G(j\omega)P(j\omega) \\ S(j\omega) &= C(j\omega)P(j\omega) \\ S_o(j\omega) &= H(j\omega)S(j\omega) \end{aligned} \tag{3.8}$$

and the auto power spectral densities are (Carlson 1975; Stanley 1982)

$$\begin{aligned}
G_{pp}(\omega) &= |P(j\omega)|^2 \\
G_{ss}(\omega) &= |S(j\omega)|^2 = \alpha_s G_{pp}(\omega) \\
G_{sso}(\omega) &= |S_o(j\omega)|^2 \\
G_{ppo}(\omega) &= |P_o(j\omega)|^2 \\
G_{cco}(\omega) &= |P_o(j\omega) + S_o(j\omega)|^2
\end{aligned} \tag{3.9}$$

where α_s denotes the ratio of the auto power spectral densities of the secondary wave to the primary wave, or simply the ratio of the secondary wave power to the primary wave power, before propagation:

$$\alpha_s = \frac{G_{ss}(\omega)}{G_{pp}(\omega)}. \tag{3.10}$$

Using equations (3.4), (3.5), (3.8), and (3.9) the spectral densities of $p_o(t)$ and $s_o(t)$ can further be written as

$$\begin{aligned}
G_{ppo}(\omega) &= \frac{A^2}{r_g^2} G_{pp}(\omega) \\
G_{sso}(\omega) &= \alpha_s \frac{A^2}{r_h^2} G_{pp}(\omega)
\end{aligned} \tag{3.11}$$

and the spectral density of $o(t)$ as

$$G_{cco}(\omega) = |G(j\omega)P(j\omega) + H(j\omega)C(j\omega)P(j\omega)|^2.$$

This, after simplification, yields

$$G_{cco}(\omega) = G_{ppo}(\omega) + G_{sso}(\omega) + 2\sqrt{G_{ppo}(\omega)G_{sso}(\omega)} \cos\left[\omega\frac{\Delta r}{c} + \theta_c(\omega)\right] \tag{3.12}$$

where $\Delta r = r_g - r_h$.

Substituting for $G_{cco}(\omega)$ from equation (3.12) into equation (3.3) and simplifying yields

$$K = -\frac{G_{sso}(\omega)}{G_{ppo}(\omega)} - 2\sqrt{\frac{G_{sso}(\omega)}{G_{ppo}(\omega)}} \cos\left[\omega\frac{\Delta r}{c} + \theta_c(\omega)\right]. \tag{3.13}$$

If the ratio of the spectral densities of the secondary wave to the primary wave, or simply, the ratio of the secondary wave power to the primary wave power at the observation point is denoted by α and the cross-spectral density factor defined by the cosinusoidal factor in equations (3.12) and (3.13) is denoted by β:

$$\alpha = \frac{G_{\mathrm{sso}}(\omega)}{G_{\mathrm{ppo}}(\omega)}$$

$$\beta = \cos\left[\omega\frac{\Delta r}{c} + \theta_c(\omega)\right] \tag{3.14}$$

then equation (3.13) can be written as

$$K = -2\beta\sqrt{\alpha} - \alpha. \tag{3.15}$$

This gives an analytical relationship between the cross-power spectral density factor, interpreted as the relative phase of the component waves, the auto power spectral density ratio, interpreted as the relative amplitude of the component waves, and the degree of cancellation given by the cancellation factor K.

3.3 General conditions for cancellation

The general conditions necessary for field cancellation are developed in this section. These provide the basis for the synthesis of ANC systems.

It follows from equations (3.1) and (3.3) that, for cancellation to occur, the field cancellation factor must be between zero and unity:

$$0 < K \leq 1. \tag{3.16}$$

A zero value for K corresponds to no cancellation and a unity value of K corresponds to complete cancellation.

Substituting for K from equation (3.15) into equation (3.16) and simplifying yields

$$\frac{\sqrt{\alpha}}{2} < -\beta \leq \frac{1+\alpha}{2\sqrt{\alpha}}. \tag{3.17}$$

Since α is the ratio of powers of the waves it is therefore a positive real number. This implies that both the left- and right-hand sides of equation (3.17) are positive. This can hold only if the cross-spectral density factor assumes negative values:

$$\beta < 0 \tag{3.18}$$

This is the phase condition for cancellation.

As given by equation (3.17) $-\beta$ falls between the functions $\zeta = \sqrt{\alpha}/2$ and $\gamma = (1+\alpha)/2\sqrt{\alpha}$. This is shown in Fig. 3.3. However, it is known from the

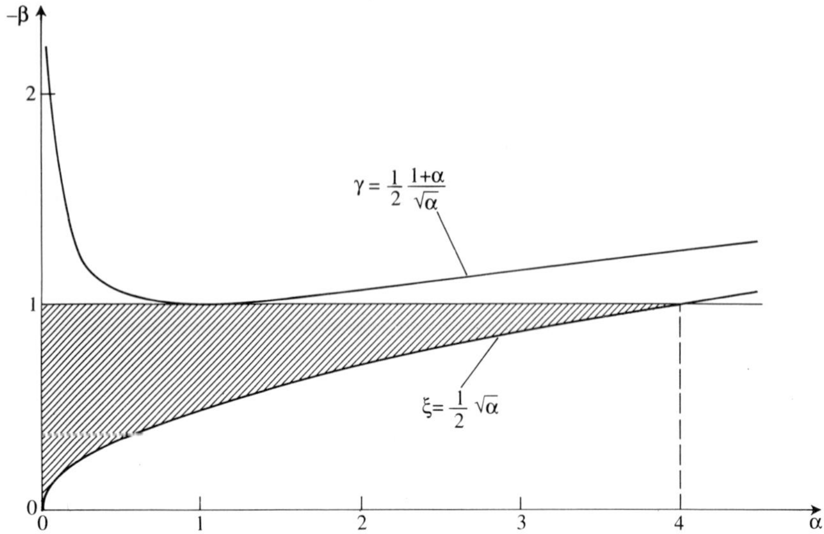

Fig. 3.3. Amplitude and phase conditions of cancellation

definition of β that the cross-spectral density factor is bounded in magnitude by unity:

$$|\beta| \leq 1. \tag{3.19}$$

Therefore, equation (3.17) becomes

$$\frac{\sqrt{\alpha}}{2} < -\beta \leq 1. \tag{3.20}$$

Thus, the required amplitude condition for cancellation is given by

$$0 < \frac{\sqrt{\alpha}}{2} < -\beta. \tag{3.21}$$

The combined amplitude and phase conditions of cancellation is indicated by the shaded region in Fig. 3.3.

Squaring the inequality in equation (3.21) yields the range of permissible power ratio, for cancellation to be feasible, as

$$0 < \alpha < 4\beta^2. \tag{3.22}$$

Substituting for α from equation (3.14) into equation (3.22) yields

$$0 < G_{\text{sso}}(\omega) < 4\beta^2 G_{\text{ppo}}(\omega). \tag{3.23}$$

Note that if $-\beta$ is at its upper limit (unity) then equations (3.22) and (3.23) imply that cancellation at the observation point can occur if the spectral density of the secondary wave is less than four times the spectral density of the primary wave or, equivalently, if the power associated with the secondary wave does not exceed four times the power associated with the primary wave.

Substituting for β from equations (3.14) into equation (3.20) yields

$$\frac{\sqrt{\alpha}}{2} < -\cos\left[\omega\tau + \theta_c(\omega)\right] < 1; \qquad 0 < \alpha < 4 \tag{3.24}$$

where τ denotes the relative time delay between primary and secondary waves, of the same frequency, reaching the observation point:

$$\tau = \frac{\Delta r}{c} = t_g - t_h. \tag{3.25}$$

It follows from equation (3.24) that for cancellation to occur at the observation point the phase difference $\omega\tau + \theta_c(\omega)$ between the same frequency components of the primary and secondary waves must satisfy the following relation

$$(2n+1)\pi - \cos^{-1}\frac{\sqrt{\alpha}}{2} < \omega\tau + \theta_c(\omega) < (2n+1)\pi + \cos^{-1}\frac{\sqrt{\alpha}}{2}$$

$$\text{for } n = 0, \pm 1, \pm 2, \ldots \text{and } 0 < \alpha < 4. \tag{3.26}$$

This is shown in Fig. 3.4. Note that the phase difference is measured here as a positive angle.

The relations developed above are the general conditions necessary for field cancellation. These are given in terms of the relative amplitude and relative phase of the component waves upon which the synthesis of ANC systems can be based.

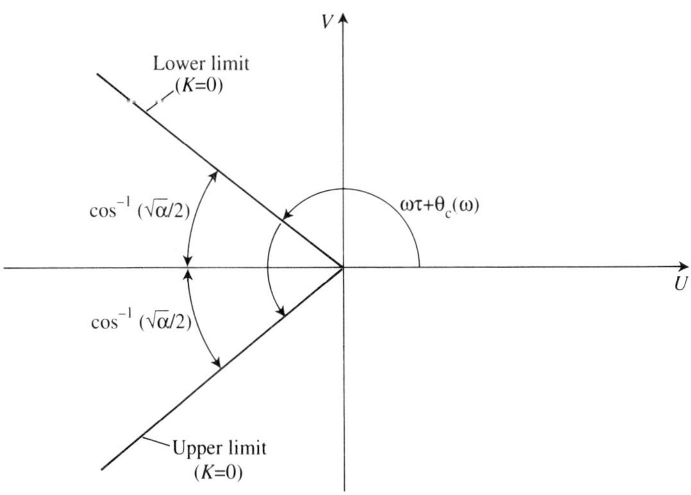

Fig. 3.4. Range of difference in phase between component waves for cancellation

3.4 Three-dimensional description of cancellation

The cancellation factor, as noted in the previous section, is a function of the power ratio α and the cross-spectral density factor β. These in turn are functions of the frequency ω of the component waves and relative distance from the sources to the observation point. In the first part of this section a spatial description of these functions is obtained. These are then used to map the conditions of cancellation on to the system geometry. Finally, a geometric description of the pattern of cancellation and reinforcement in the three-dimensional propagation medium is obtained.

3.4.1 Power ratio and cross-spectral density factor

Consider Fig. 3.5 which describes Fig. 3.1 in the three-dimensional *UVW*-space, with points *P*, *S*, and *O*, respectively, representing the locations of the primary source, secondary source, and observer. The distances r_g and r_h are thus given by

$$r_g = \left[\left(u - \frac{d}{2}\right)^2 + v^2 + w^2\right]^{1/2}$$

$$r_h = \left[\left(u + \frac{d}{2}\right)^2 + v^2 + w^2\right]^{1/2}. \tag{3.27}$$

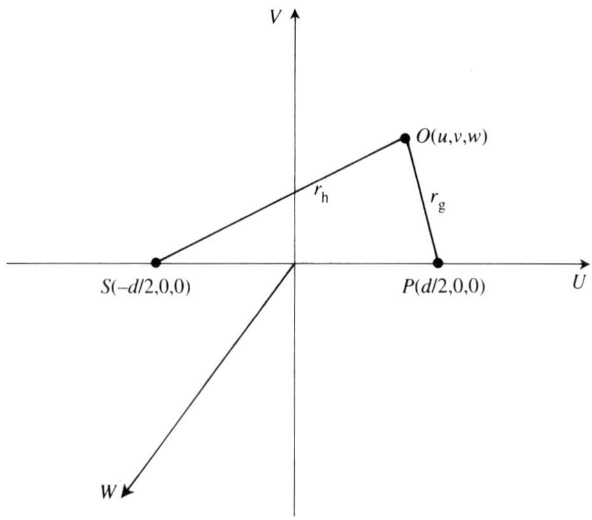

Fig. 3.5. Primary and secondary sources in three-dimensional coordinates

Substituting for $G_{\text{ppo}}(\omega)$ and $G_{\text{sso}}(\omega)$ from equations (3.11) into equation (3.14) and simplifying yields

$$\alpha = a^2 \alpha_s \tag{3.28}$$

where a is a positive real number defined by

$$a = \frac{r_g}{r_h}. \tag{3.29}$$

It follows from equation (3.28) that, provided α_s is constant, the variation of the power ratio α as a function of the location of the observation point in the three-dimensional space is described by the distance ratio a.

Substituting for r_g and r_h from equations (3.27) into equation (3.29) and simplifying yields (see Appendix A)

$$\left[u - \left(\frac{1 + a^2}{1 - a^2}\right)\frac{d}{2}\right]^2 + v^2 + w^2 = \left[\frac{ad}{1 - a^2}\right]^2. \tag{3.30}$$

As the distance ratio a varies between zero (location of the primary source) to infinity (location of the secondary source) equation (3.30) describes a family of spheres with centres on the U-axis with u-coordinates of $(1 + a^2)/(1 - a^2)d/2$ and radii $|ad|/|1 - a^2|$. This family is shown in Fig. 3.6 in the two-dimensional UV-plane, for $d = 4$ units, revolution of which around the U-axis gives the corresponding three-dimensional spherical

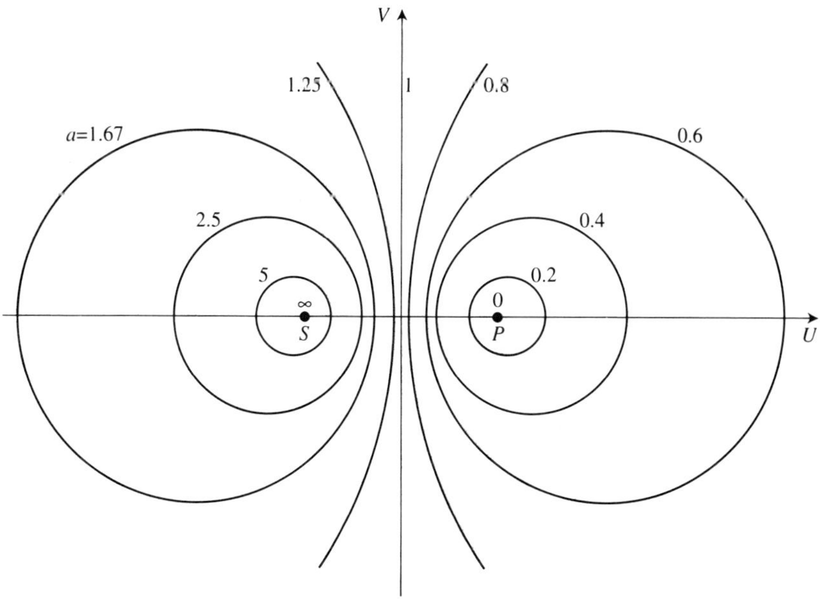

Fig. 3.6. Loci of constant distance ratios for $d = 4$

surfaces. Note that for $a = 1$ the spherical surface changes to a plane surface coinciding with the VW-plane. Each surface in equation (3.30) thus represents a certain distance ratio a and, for constant α_s, corresponds to a certain power ratio α. Therefore, equation (3.30) is referred to as loci of constant distance ratios or, equivalently, as loci of constant power ratios.

It follows from equation (3.14) that for a single tone of the component waves the cross-spectral density factor β is a function of the distance difference Δr only. This implies that a description of β in the three-dimensional space is given by the distance difference Δr. Let Δr be denoted by a real number b:

$$\Delta r = r_g - r_h = b; \qquad |b| \leq d. \tag{3.31}$$

Substituting for r_g and r_h from equations (3.27) into equation (3.31) and simplifying yields (see Appendix B)

$$\frac{u^2}{\left(\dfrac{b^2}{4}\right)} - \frac{v^2}{\left(\dfrac{d^2 - b^2}{4}\right)} - \frac{w^2}{\left(\dfrac{d^2 - b^2}{4}\right)} = 1. \tag{3.32}$$

As the distance difference b varies from $-d$ to $+d$, equation (3.32) defines a family of hyperbolic surfaces with foci coinciding either with point P or point S. This family is shown in Fig. 3.7 in the two-dimensional UV-plane, revolution of which around the U-axis yields the corresponding family of three-dimensional hyperbolic surfaces. Note that for $b = 0$ the hyperbolic surface changes to a plane surface coinciding with the VW-plane. Each surface in equation (3.32) thus represents a certain distance difference b or, equivalently, a certain value of the cross-spectral density factor β. Therefore, equation (3.32) is referred to as loci of constant distance differences or, equivalently, for single tones of the component waves, as loci of constant cross-spectral density factors. Note that the distance difference Δr is directly related to the relative time delay τ between the primary and secondary waves reaching the observation point O; see equation (3.25). Therefore, equation (3.32) can also be referred to as the loci of constant time differences τ.

The locus of points O for which both the distance ratio a and the distance difference b are constant is described by the intersection of the surfaces in equations (3.30) and (3.32); see Appendix C. Such an intersection will exist if the condition

$$\frac{|b|}{d}\left(\frac{1 + a}{1 - a}\right) \geq 1, \qquad \text{for } a < 1$$

$$\tag{3.33}$$

$$\frac{|b|}{d}\left(\frac{a + 1}{a - 1}\right) \geq 1, \qquad \text{for } a > 1$$

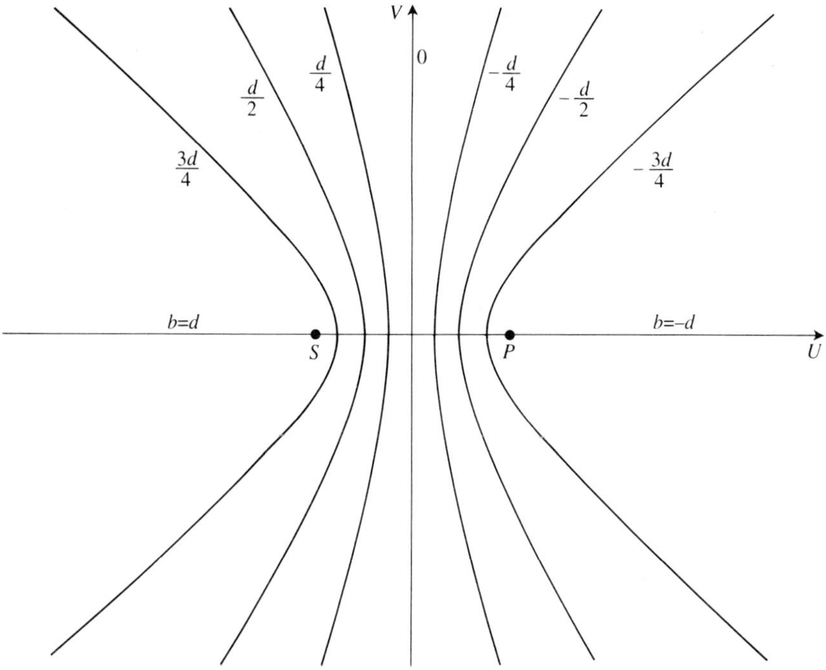

Fig. 3.7. Loci of constant distance differences

holds. Under this, the intersection of the two surfaces will form a circle in the three dimensional space of Fig. 3.5. The circle thus formed will be on a plane parallel to the VW-plane, with centre on the U-axis with u- coordinate given by u_c, where

$$u_c = \frac{b^2}{2d}\left(\frac{1+a}{1-a}\right). \tag{3.34}$$

In such a manner the circle is defined by

$$v^2 + w^2 = \frac{d^2}{4}\left[1 - \left(\frac{b}{d}\right)^2\right]\left[\left(\frac{b}{d}\right)^2\left(\frac{1+a}{1-a}\right)^2 - 1\right]; \qquad a \neq 1. \tag{3.35}$$

Note that for $a = 1$ the distance difference b must be zero for the intersection to be possible:

$$b = 0; \qquad a = 1. \tag{3.36}$$

This defines a plane surface that coincides with the VW-plane.

Equation (3.35), for sets of a and b, defines a family of circles and equation (3.36) defines a plane surface. Both equations (3.35) and (3.36) define loci for which both the distance ratio a and distance difference b are constant.

Therefore, these are referred to as loci of constant a–constant b or, equivalently, for single tones of the component waves, as loci of constant α–constant β.

Having found a three-dimensional description of the basic parameters relating the amplitudes and phases of the component waves, attention is now focused on mapping the conditions of cancellation on to the system geometry.

3.4.2 *Cancellation as a function of power ratio*

Treating β as a parameter in equation (3.15), that is, restricting the observation point to constant β surfaces, and varying the point so as to coincide with various distance ratios or, equivalently, to define various α surfaces, will result in the cancellation factor K as a function of α forming the family of curves shown in Fig. 3.8(a) and (b). Note that Fig. 3.8(a) corresponds to cancellation and Fig. 3.8(b) corresponds to reinforcement. Each curve in Fig. 3.8 shows cancellation as a function of the power ratio α on the corresponding constant β surface.

It follows from equation (3.22) that on a β surface, with $\beta < 0$, cancellation occurs at points which are between α surfaces corresponding to power ratios of zero and $4\beta^2$.

Maximizing the field cancellation factor in equation (3.15) with respect to the power ratio α it follows that maximum cancellation is obtained at such points on a β surface, $\beta < 0$, for which the following relation holds

$$\alpha = \beta^2 \tag{3.37}$$

corresponding to which the maximum value of the field cancellation factor, $K_{\max}$, is

$$K_{\max} = \beta^2. \tag{3.38}$$

It follows from equations (3.37) and (3.38) that the maxima of the family of curves in equation (3.15) form a straight line;

$$K = \alpha \qquad \text{for } 0 < \alpha \leq 1. \tag{3.39}$$

This line is shown in Fig. 3.8.

Solving equation (3.15) for $-\beta$ and substituting into equation (3.20), after simplification, yields

$$0 < K \leq 2\sqrt{\alpha} - \alpha. \tag{3.40}$$

Solution of the right-hand side of this inequality for α in terms of K reveals that cancellation larger than that corresponding to any K is possible for waves for which the power ratio α is within the limits

$$\left[1 - \sqrt{1 - K}\right]^2 \leq \alpha \leq \left[1 + \sqrt{1 - K}\right]^2 \tag{3.41}$$

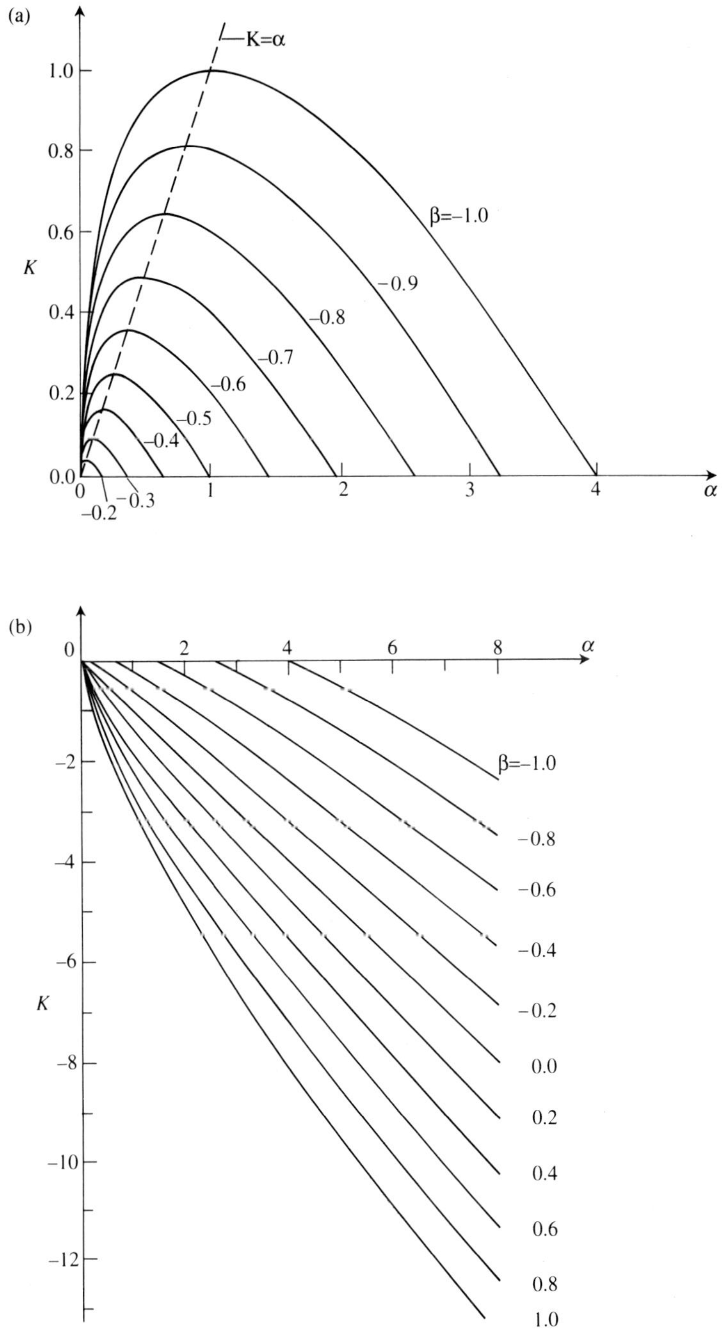

Fig. 3.8. Field cancellation factor as function of power ratio: (a) $K > 0$; (b) $K < 0$

while the magnitude of β should be higher than $|\beta|_{min}$, where $|\beta|_{min}$, from a maximization of equation (3.15) for β as a function of α, is

$$|\beta|_{min} = \sqrt{K}. \tag{3.42}$$

It follows from Fig. 3.8(b) that on β surfaces corresponding to $\beta > 0$ the cancellation factor K is always negative implying that under such a situation the primary wave is always reinforced. As seen, the reinforcement on these surfaces increases with increasing power ratios.

Example 3.1 Let the separation d between the sources in Fig. 3.5 be 1 metre and the distance difference b be -0.6 metres. Substituting for b and d into equation (3.32) and simplifying yields

$$\frac{u^2}{0.09} - \frac{v^2}{0.16} - \frac{w^2}{0.16} = 1. \tag{3.43}$$

This is a hyperbolic surface corresponding to $b = -0.6$ metres.

If the sources emit a single tone of 100 Hz with a source power ratio $\alpha_s = 1$ and initial phase difference $\theta_c(\omega) = 180°$ then using equation (3.14), with $c = 340$ m/sec, the value of β corresponding to the surface in equation (3.43) is $\beta = -0.4457$.

Substituting for β into equation (3.22) yields the range of permissible power ratio for cancellation as $0 < \alpha < 0.7947$. Using equation (3.28) the corresponding range of distance ratio is obtained as $0 < a < 0.8915$. This represents the region of the three-dimensional UVW-space that is between point P and the sphere in equation (3.30) for $a = 0.8915$ and $d = 1$:

$$(u - 4.3709)^2 + v^2 + w^2 = 4.3422^2. \tag{3.44}$$

The hyperbolic surface in equation (3.43) is shown in two dimensions in Fig. 3.9, revolution of which around the U-axis results in the corresponding three-dimensional picture. The range of cancellation is indicated by the dashed-line portions of the surface which includes the base of the surface from $u = 0.3$ to $u = 3.128$ metres. The solid-line portion of the curve corresponds to regions of reinforcement on which reinforcement is minimum at points with $u = 3.128$ metres and increases as the point moves away from the origin. Maximum cancellation occurs at points on the surface (a circular strip) for which equation (3.39) holds. Using equations (3.37) and (3.28) these points are found to be common to the surface in equation (3.43) and the constant α surface of equation (3.30) with $\alpha = 0.1987$ or, equivalently, $a = 0.4458$, $d = 1$ metres:

$$(u - 0.7489)^2 + v^2 + w^2 = 0.5563^2. \tag{3.45}$$

These points form a circular strip around the U-axis at $u = 0.4696$ metres.

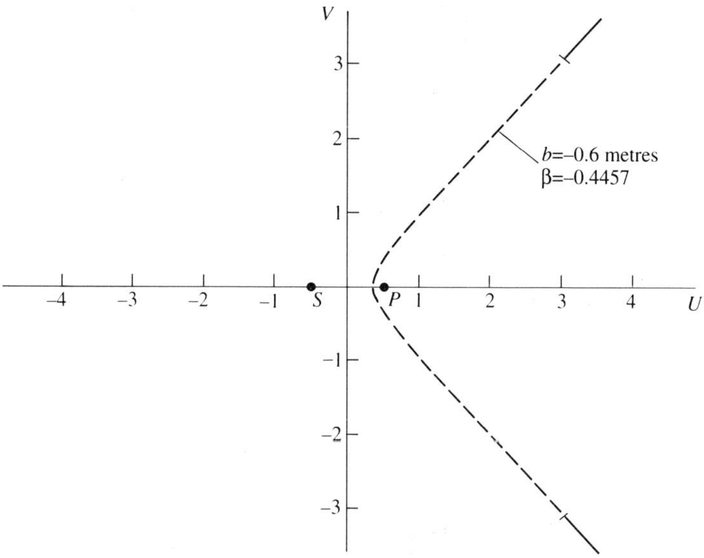

Fig. 3.9. Cancellation and reinforcement pattern on a $b = -0.6$ metres surface

3.4.3 Cancellation as a function of distance difference

Treating α as a parameter in equation (3.15), that is, restricting the observation point to constant α surfaces, and varying the point so as to coincide with various Δr surfaces or, equivalently, define various β surfaces, will result in the cancellation factor K as a function of β forming the family of straight lines shown in Fig. 3.10. As β varies between -1 and $+1$ the cancellation factor K remains less than or equal to unity. For cancellation to occur at points on an α surface the corresponding power ratio α should be within 0 and 4:

$$0 < \alpha < 4 \tag{3.46}$$

while the corresponding value of β should remain within the range

$$-1 \leq \beta < -\frac{\sqrt{\alpha}}{2}. \tag{3.47}$$

It follows from the above that on a constant α surface satisfying equation (3.46) the process of interference will correspond to a combined pattern of cancellation and reinforcement in a manner that maximum cancellation will occur at points for which $\beta = -1$. On either side of these points cancellation will decrease linearly as the observation point varies in the direction of increasing β. There will be neither cancellation nor reinforcement ($K = 0$) at points for which $\beta = \sqrt{\alpha}/2$. Beyond these points, in the direction of increasing β, the situation will correspond to reinforcing the primary wave. Maximum reinforcement will occur at points for which β is $+1$.

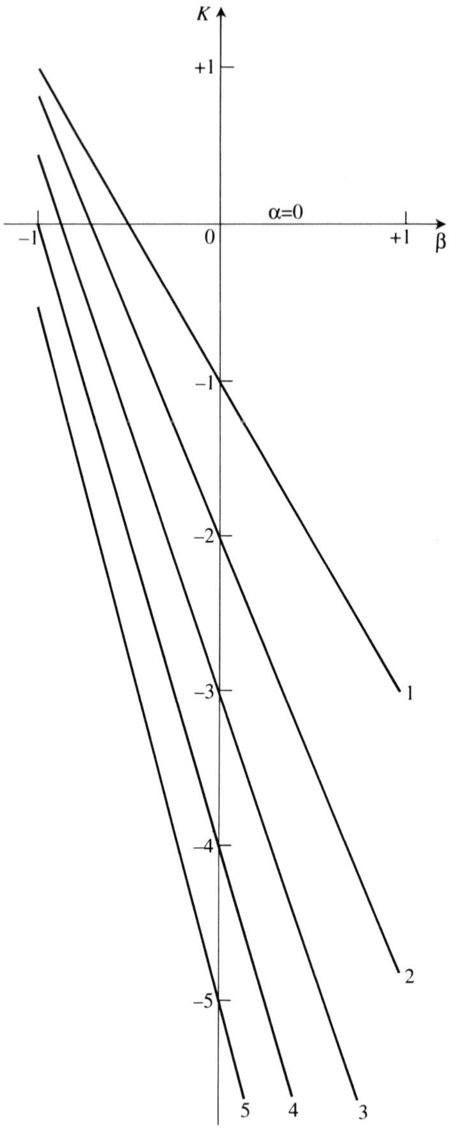

Fig. 3.10. Field cancellation factor as function of cross-spectral density factor

On a constant α surface for which α is outside the limits given in equation (3.46) the primary wave is reinforced at all points of the surface. Here, reinforcement varies linearly with β in such a manner that it is minimum at points for which $\beta = -1$ and maximum at points for which $\beta = +1$.

Substituting for τ from equation (3.25) into equation (3.26), using equation (3.6), and simplifying yields

$$\left[(2n+1) - \frac{1}{\pi}\theta_c(\omega) - \frac{1}{\pi}\cos^{-1}\frac{\sqrt{\alpha}}{2}\right]\frac{\lambda}{2} < \Delta r < \left[(2n+1) - \frac{1}{\pi}\theta_c(\omega) + \frac{1}{\pi}\cos^{-1}\frac{\sqrt{\alpha}}{2}\right]\frac{\lambda}{2} \tag{3.48}$$

for $n = 0, \pm 1, \pm 2, \ldots$ and $0 < \alpha < 4$.

This implies that, on a constant α surface, the range of cancellation is directly proportional to the wavelength or, in other words, inversely proportional to the frequency of the component waves.

If the range of cancellation given in equation (3.48) is to cover all points of the constant α surface then it should be greater than the maximum value $|b|$ can have:

$$\text{Range of cancellation} \geq d \tag{3.49}$$

or, using equations (3.6) and (3.48) for $n = 0$, the following should hold

$$\left|\frac{\omega d}{c} + \theta_c(\omega) - \pi\right| < \cos^{-1}\frac{\sqrt{\alpha}}{2}; \qquad 0 < \alpha < 4. \tag{3.50}$$

As a special case, if $\theta_c(\omega) = 180°$, then equation (3.50) yields

$$d < \frac{c}{\omega}\cos^{-1}\frac{\sqrt{\alpha}}{2}; \qquad 0 < \alpha < 4. \tag{3.51}$$

This is an inverse relation between source separation and source frequency. As an example, to ensure cancellation of a 100 Hz tone at all points of a constant α surface of $\alpha = 0.4096$ the source separation must not exceed 0.6737 metres, whereas if the source frequency is 500 Hz then the upper limit of d will be 0.1347 metres.

Substituting for β from equations (3.14) into equation (3.15) and using equation (3.6) yields

$$K = -2\sqrt{\alpha}\cos\left[\frac{\Delta r}{\lambda/2}\pi + \theta_c(\omega)\right] - \alpha. \tag{3.52}$$

It follows from this relation that cancellation on a constant α surface ($0 < \alpha < 4$) is a cosinusoidal function of the distance difference Δr with a period equal to the signal wavelength. The cancellation factor has an average value of $-\alpha$ implying that for non-zero α the range of cancellation is always less than the range of reinforcement. Within the cancellation range, given by equation (3.48), cancellation reaches its maximum at such Δr surfaces for which $\beta = -1$:

$$\cos\left[\frac{\Delta r}{\lambda/2}\pi + \theta_c(\omega)\right] = -1$$

or

$$\Delta r = \left[(2n+1) - \frac{1}{\pi}\theta_c(\omega)\right]\frac{\lambda}{2}; \qquad n = 0, \pm 1, \pm 2, \ldots. \tag{3.53}$$

Therefore, on a constant α surface $(0 < \alpha < 4)$, maximum cancellation occurs at such Δr surfaces for which the distance difference is directly proportional to half the wavelength. Note that if $\theta_c(\omega)$ in equation (3.53) is zero then maximum cancellation occurs at those Δr surfaces for which the distance difference is an odd multiple of half the wavelength, whereas if $\theta_c(\omega) = 180°$ then maximum cancellation occurs at distance differences that are even multiples of half the wavelength.

If follows from equation (3.48) that cancellation will begin to diminish $(K = 0)$ on either side of the Δr surface defined by equation (3.53) at surfaces $b_1 = \Delta r_1$ and $b_2 = \Delta r_2$, where

$$b_1 = \left[2n + 1 - \frac{1}{\pi}\theta_c(\omega)\right]\frac{\lambda}{2} - \frac{\lambda}{2\pi}\cos^{-1}\frac{\sqrt{\alpha}}{2}; \qquad n = 0, \pm 1, \pm 2, \ldots$$

$$b_2 = \left[2n + 1 - \frac{1}{\pi}\theta_c(\omega)\right]\frac{\lambda}{2} + \frac{\lambda}{2\pi}\cos^{-1}\frac{\sqrt{\alpha}}{2}; \qquad n = 0, \pm 1, \pm 2, \ldots. \tag{3.54}$$

Thus, from equations (3.52) and (3.54), the width (range) of a cancellation zone R_c and the width (range) of a reinforcement zone R_r are given by

$$R_c = b_2 - b_1 = \frac{\lambda}{\pi}\cos^{-1}\frac{\sqrt{\alpha}}{2}; \qquad 0 < \alpha < 4$$

$$R_r = \lambda - R_c = \lambda\left[1 - \frac{1}{\pi}\cos^{-1}\frac{\sqrt{\alpha}}{2}\right]; \qquad 0 < \alpha < 4. \tag{3.55}$$

Note that if $\alpha = 0$ (location of the primary source) then both R_c and R_r are equal to half the signal wavelength whereas if $\alpha \neq 0$ then R_c is less than R_r. Fig. 3.11 shows R_c and R_r as functions of the power ratio α of the component waves.

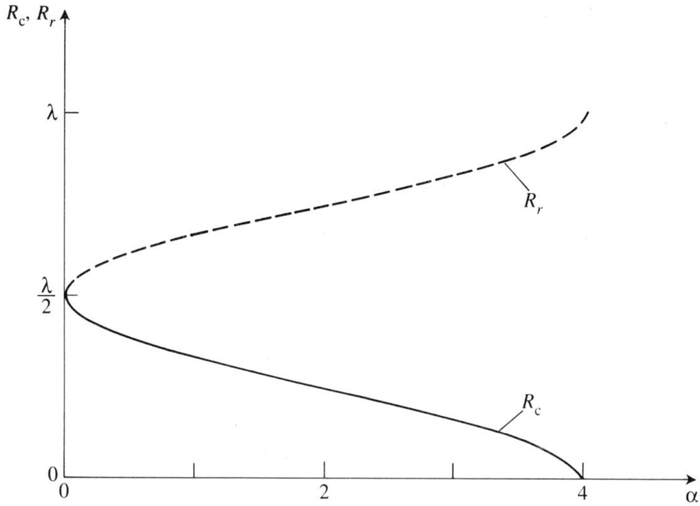

3.11. Ranges of cancellation and reinforcement zones for a single tone of component waves as function of power ratio

It follows from equation (3.55) that, on a constant α surface ($0 < \alpha < 4$), the range of cancellation R_c is the change in distance difference $|\Delta b|$ between two successive β surfaces corresponding to $K = 0$ (or $\beta = -\sqrt{\alpha}/2$) provided maximum cancellation ($\beta = -1$) occurs between the two surfaces. Likewise R_r is the change in distance difference between two successive β surfaces corresponding to $K = 0$ provided maximum reinforcement ($\beta = 1$) occurs between the two surfaces. This in relation to Fig. 3.10 means that to measure R_c then β must vary from $-\sqrt{\alpha}/2$ to -1 and further to $-\sqrt{\alpha}/2$ and, likewise, to measure R_r then β must vary from $-\sqrt{\alpha}/2$ to $+1$ and further to $-\sqrt{\alpha}/2$.

Example 3.2 Consider the primary and secondary sources in Fig. 3.5 with $d = 1$ metres, $\theta_c(\omega) = 180°$, and $\alpha_s = 1$ at a frequency of 500 Hz ($\lambda = 0.68$ metres, $c = 340$ m/sec).

Let the observer be restricted to such points for which the distance ratio is constant at $a = 0.8$. The locus of such points, defined by equation (3.30), is thus

$$(u - 2.2778)^2 + v^2 + w^2 = 2.2222^2 \qquad (3.56)$$

corresponding to which the power ratio, using equation (3.28), is $\alpha = 0.64$. This is shown in two dimensions in Fig. 3.12, revolution of which around the U-axis results in the corresponding three-dimensional surface.

Points of the limit of cancellation (LC), $K = 0$, on the surface in equation (3.56) are given by the intersection of the sphere and the hyperbolic surfaces corresponding to those in equation (3.54):

$$b_1 = 0.68n - 0.1255; \qquad n = 0, -1$$

$$b_2 = 0.68n + 0.1255; \qquad n = -1.$$

These points are marked as A, A', B, B', F, and F' in Fig. 3.12. Points of maximum cancellation (MC), using equation (3.54), are given by the intersection of equation (3.56) and the hyperbolic surface defined by equation (3.53). This corresponds to $b = -0.68$. These points are marked as M and M' in Fig. 3.12. The points of maximum reinforcement (MR) are marked as L and L' in Fig. 3.12. These are given by the intersection of equation (3.56) and the hyperbolic surface for $\beta = +1$. Using equation (3.14) the corresponding distance difference is $b = -\lambda/2 = -0.34$ metres. These results, along with the coordinates of points and corresponding values of the cancellation factor, are listed in Table 3.1.

In the two-dimensional plot of Fig. 3.12 the dashed lines indicate regions of cancellation and the solid lines represent regions of reinforcement. Revolving this around the U-axis yields the corresponding spherical surface for $\alpha = 0.64$ on which regions of cancellation are strips of the surface from $u = 0.0556$ metres to $u = 0.0709$ metres and from $u = 1.3836$ metres to $u = 2.9200$

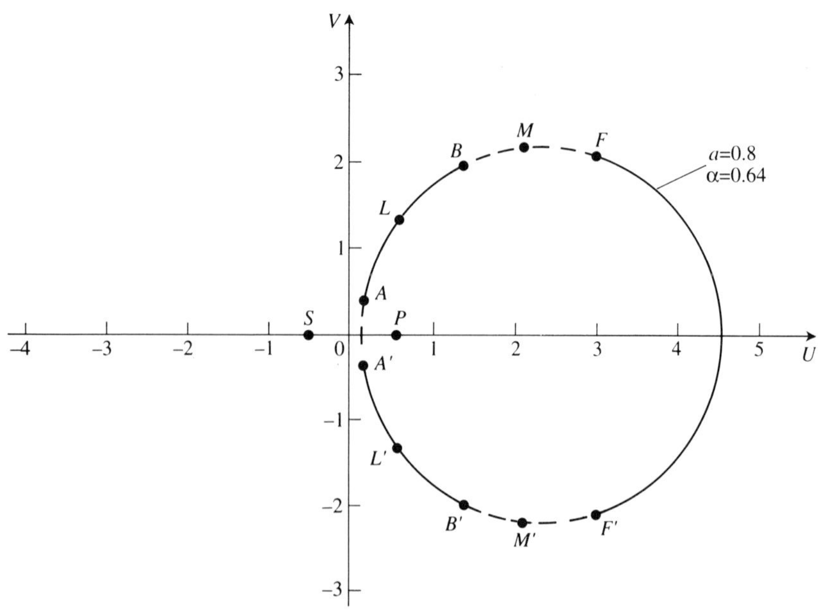

3.12. Cancellation and reinforcement pattern on an $a = 0.8$ surface

TABLE 3.1 Points of limit of cancellation (LC), maximum cancellation (MC) and maximum reinforcement (MR) in association with Fig. 3.12.

Points	Indication	Coordinates u (metres)	v (metres)	b (metres)	β	K
A	LC	+0.0709	+0.2603	−0.1255	−0.4000	0.0000
A'	LC	+0.0709	−0.2603	−0.5534	−0.4000	0.0000
B	LC	+1.3836	+2.0343	−0.5545	−0.4000	0.0000
B'	LC	+1.3836	−2.0343	−0.5545	−0.4000	0.0000
F	LC	+2.9200	+2.1274	−0.8055	−0.4000	0.0000
F'	LC	+2.9200	−2.1274	0.8055	−0.40000	0.0000
L	MR	+0.5202	+1.3598	−0.3400	+1.0000	−2.2400
L'	MR	+0.5202	−1.3598	−0.3400	+1.0000	−2.2400
M	MC	+2.0808	+2.2135	−0.6800	−1.0000	+0.9600
M'	MC	+2.0808	−2.2135	−0.6800	−1.0000	+0.9600

metres. The rest of the surface will correspond to regions of reinforcement. At points of the limit of cancellation the cross-spectral density factor β is $-\sqrt{\alpha/2} = -0.40$ as expected.

The extent of zones of cancellation and reinforcement (R_c and R_r), as defined by equations (3.55), will correspond respectively to $|\Delta b|$ values from point B to F or B' to F' for cancellation, and from point A to B or A' to B' for reinforcement. Thus, using Table 3.1

$$R_c = |(b \text{ at } B) - (b \text{ at } F)|$$

$$R_r = |(b \text{ at } A) - (b \text{ at } B)|$$

or

$$R_c = |-0.5545 + 0.8055| = 0.251 \text{ metres}$$

$$R_r = |-0.1255 + 0.5545| = 0.429 \text{ metres.}$$

Alternatively, using equations (3.55) yields

$$R_c = 0.369\,\lambda$$

$$R_r = 0.631\,\lambda.$$

Substituting for $\lambda = 0.68$ metres yields $R_c = 0.251$ metres and $R_r = 0.429$ metres which agree with the values obtained using Table 3.1. It follows from this analysis that an indirect method of measuring the wavelength of the source signal will be to add R_c and R_r, as determined from measurements of Δr at points of the limit of cancellation on a constant α surface in the direction of changing β.

3.4.4 Cancellation as a function of frequency

To study cancellation as function of frequency both the power ratio α and distance difference Δr should be constant. This, in terms of a three-dimensional description, means that the observer should be restricted to points defining constant a-constant b circle of equation (3.35) or the surface in equation (3.36).

Substituting for β from equations (3.14) into equation (3.15) yields

$$K = -2\sqrt{\alpha}\cos\left[\frac{\Delta r}{c}\omega + \theta_c(\omega)\right] - \alpha \tag{3.57}$$

or

$$K = -2\sqrt{\alpha}\cos\left[\theta(\omega)\right] - \alpha \tag{3.58}$$

where $\theta(\omega) = \Delta r\omega/c + \theta_c(\omega)$. This implies that cancellation of a broadband noise varies as a cosine function of $\theta(\omega)$. Moreover, if the power ratio α is within 0 and 4 over a broad frequency range then a cancellation/reinforcement pattern will result. However, if the power ratio exceeds 4 then the situation will correspond to reinforcing the primary wave.

Manipulation of equation (3.58) reveals that, at observation points for which the power ratio is within 0 and 4, maximum cancellation occurs at frequencies ω_m for which

$$\cos\left[\theta(\omega_m)\right] = -1$$

or

$$\theta(\omega_m) = (2n+1)\pi; \qquad n = 0, \pm 1, \pm 2, \ldots. \tag{3.59}$$

Furthermore, cancellation will reach its lowest limit ($K = 0$) on either side of ω_m at frequencies ω_1, where

$$\theta(\omega_1) = (2n+1)\pi \pm \cos^{-1}\frac{\sqrt{\alpha}}{2}; \qquad 0 < \alpha < 4, n = 0, \pm 1, \pm 2, \ldots. \tag{3.60}$$

Example 3.3 As a special case of the above, consider $\theta_c(\omega) = 180°$. Substituting this into equation (3.57) yields

$$K = 2\sqrt{\alpha}\cos\left(\frac{\Delta r}{c}\omega\right) - \alpha. \tag{3.61}$$

This gives K as a cosinusoidal function of the frequency ω with period $2\pi c/\Delta r$ and an average value of $-\alpha$. Note that here $\theta(\omega) = (\Delta r/c)\omega$. Maximizing equation (3.61) with respect to ω gives the frequencies ω_m at which maximum cancellation occurs as

$$\omega_m = (2n)\frac{\pi c}{|\Delta r|}; \qquad n = 0, 1, 2, \ldots. \tag{3.62}$$

In the vicinity of ω_m cancellation will decrease as the cosine of $\Delta r\omega/c$ and will, eventually, reach its lowest limit ($K = 0$) at frequencies ω_l. Using equation (3.61) these are given by

$$\omega_l = \left[(2n)\pi \pm \cos^{-1}\frac{\sqrt{\alpha}}{2}\right]\frac{c}{|\Delta r|}; \qquad 0 < \alpha < 4, n = 0, 1, 2, \ldots. \tag{3.63}$$

Thus, if cancellation is to occur at all frequencies from zero to ω_s then ω_l in equation (3.63), with $n = 0$, must be at least equal to ω_s:

$$\omega_s \leq \omega_l = \frac{c}{|\Delta r|}\cos^{-1}\frac{\sqrt{\alpha}}{2}; \qquad 0 < \alpha < 4$$

or

$$|\Delta r| \leq \frac{c}{\omega_s}\cos^{-1}\frac{\sqrt{\alpha}}{2}; \qquad 0 < \alpha < 4. \tag{3.64}$$

For instance, to cancel noise from 0 to 500 Hz ($\omega_s = 1000\,\pi$ rad/sec) at all observation points for which $0 < \alpha < 4$ then $|\Delta r|$ must not exceed 0.113 metres.

It follows from equations (3.61) and (3.63) that the frequency band over which cancellation occurs is of width

$$B_c = \frac{2c}{|\Delta r|}\cos^{-1}\frac{\sqrt{\alpha}}{2}; \qquad 0 < \alpha < 4 \tag{3.65}$$

and the frequency band over which reinforcement occurs is of width

$$B_r = \frac{2c}{|\Delta r|}\left[\pi - \cos^{-1}\frac{\sqrt{\alpha}}{2}\right]; \qquad 0 < \alpha < 4. \tag{3.66}$$

This implies that for a non-zero α the cancellation band B_c will always be less than the reinforcement band B_r.

The theoretical results obtained above can be verified experimentally. A set of experiments and results verifying the above have been reported by Leitch and Tokhi (1987).

3.4.5 Zones of cancellation and reinforcement

It follows from the results obtained so far that a combined consideration of the cancellation and reinforcement patterns on both α and β surfaces will lead to a full pattern of zones of cancellation and reinforcement in three-dimensional space. In this manner the locus of points separating adjacent zones of cancellation and reinforcement is defined by $K = 0$. Thus, using equation (3.15), this locus is given by

$$\beta = -\frac{\sqrt{\alpha}}{2}. \tag{3.67}$$

For a point in the medium to be in a cancellation zone both conditions given in equations (3.46) and (3.47) should hold:

$$- \leq \beta < -\frac{\sqrt{\alpha}}{2}; \qquad 0 < \alpha < 4.$$

If either or both of these are not met then the point will be in a reinforcement zone. It follows from this that noise will always be reinforced within the region of medium between α surfaces of $\alpha = 4$ and $\alpha = \infty$. Outside this region, that is, between α surfaces of $\alpha = 0$ and $\alpha = 4$, the space divides into zones of cancellation and reinforcement in a manner determined by equation (3.67).

For given primary and secondary sources, there are basically two parameters; namely, separation between the sources and signal frequency that affect the extent of a cancellation/reinforcement zone. Decreasing the separation d between the sources will result in an increase in the region extending between α surfaces of $\alpha = 0$ and $\alpha = 4$; see equations (3.28) and (3.30). Furthermore, such a decrease in d will ensure that the condition in equation (3.49), and consequently equation (3.47), will hold over large volumes of the medium.

It is already evident from a consideration of cancellation as a function of frequency that the effect of signal frequency on the extent of zones of cancellation and reinforcement is similar to that of separation between the sources, that is, a decrease in the source frequency will lead to an increase in the range of cancellation.

To demonstrate the above effects consider the following two examples.

Example 3.4 Let the primary and secondary sources be located in three-dimensional space as in Fig. 3.5 and emit a single tone of 500 Hz with $\alpha_s = 1$ and $\theta_c(\omega) = 180°$. We consider the pattern of cancellation and reinforcement for $d = 1$, 0.5, and 0.1 metres.

Substituting for ω, $\theta_c(\omega)$, and $c = 340$ m/sec into equation (3.14) and for α_s into equation (3.28) yields

$$\beta = \cos(9.24\, b + \pi); \qquad b = \Delta r$$
$$\alpha = a^2 \tag{3.68}$$

where a and b, using equations (3.29), (3.31), and (3.27) are given by

$$a = \left[\frac{\left(u - \frac{d}{2}\right)^2 + v^2 + w^2}{\left(u + \frac{d}{2}\right)^2 + v^2 + w^2} \right]^{1/2} \tag{3.69}$$

$$b = \sqrt{\left(u - \frac{d}{2}\right)^2 + v^2 + w^2} - \sqrt{\left(u + \frac{d}{2}\right)^2 + v^2 + w^2}.$$

To obtain the locus of points satisfying equation (3.67) a computer program based on the following algorithm can be used.

For a given d:

(1) calculate a and b at a chosen point (u, v, w) using equation (3.69);
(2) calculate β and α using equations (3.68);
(3) test for equation (3.67): if the relation holds then store the coordinates of the point, otherwise go to (1).

The collection of points thus obtained can be used to sketch the pattern of cancellation and reinforcement. Fig. 3.13 shows such a pattern for $d = 1$, 0.5, and 0.1 metres in two dimensions. In these diagrams the shaded areas correspond to regions of reinforcement and the plain areas correspond to regions of cancellation. The axes are graduated in metres. The three-dimensional sketches corresponding to these diagrams is obtained by revolving each around the U-axis.

Example 3.5 Let the primary and secondary sources be located in three-dimensional space as in Fig. 3.5 and emit a single tone of 100 Hz with $\alpha_s = 1$ and $\theta_c(\omega) = 180°$. We consider the resultant patterns of cancellation and reinforcement for $d = 1$, 0.5, and 0.3 metres.

Substituting for ω, $\theta_c(\omega)$, and $c = 340$ m/sec into equation (3.14) and for α_s into equation (3.28) yields

$$\beta = \cos(0.924\, b + \pi); \qquad b = \Delta r$$

$$\alpha = a^2 \tag{3.70}$$

where a and b are as given in equations (3.69). A computer program based on the algorithm in Example 3.4, using equation (3.70) in place of equation (3.68), will result in a collection of points (u, v, w) for each value of d. The locus of these points, for each case, gives the pattern of cancellation and reinforcement. These patterns are shown in two dimensions in Fig. 3.14,

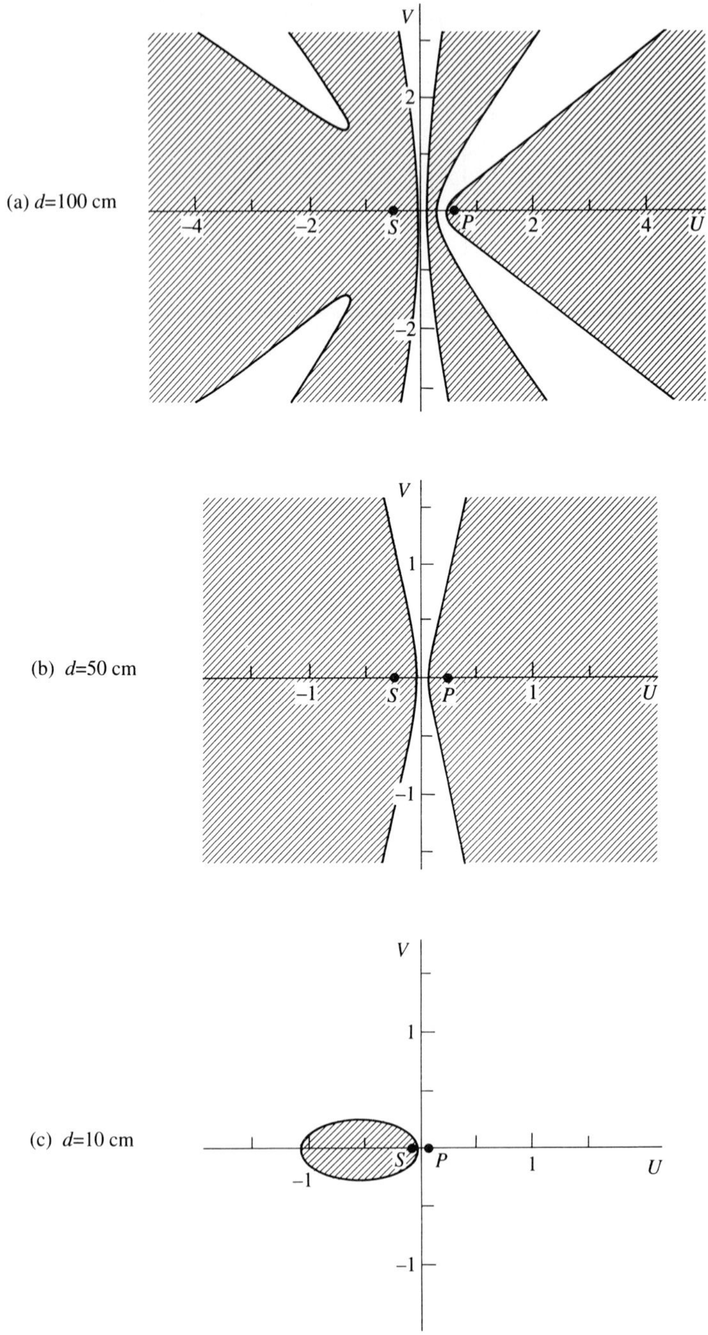

Fig. 3.13. Zones of cancellation and reinforcement (f = 500 Hz)

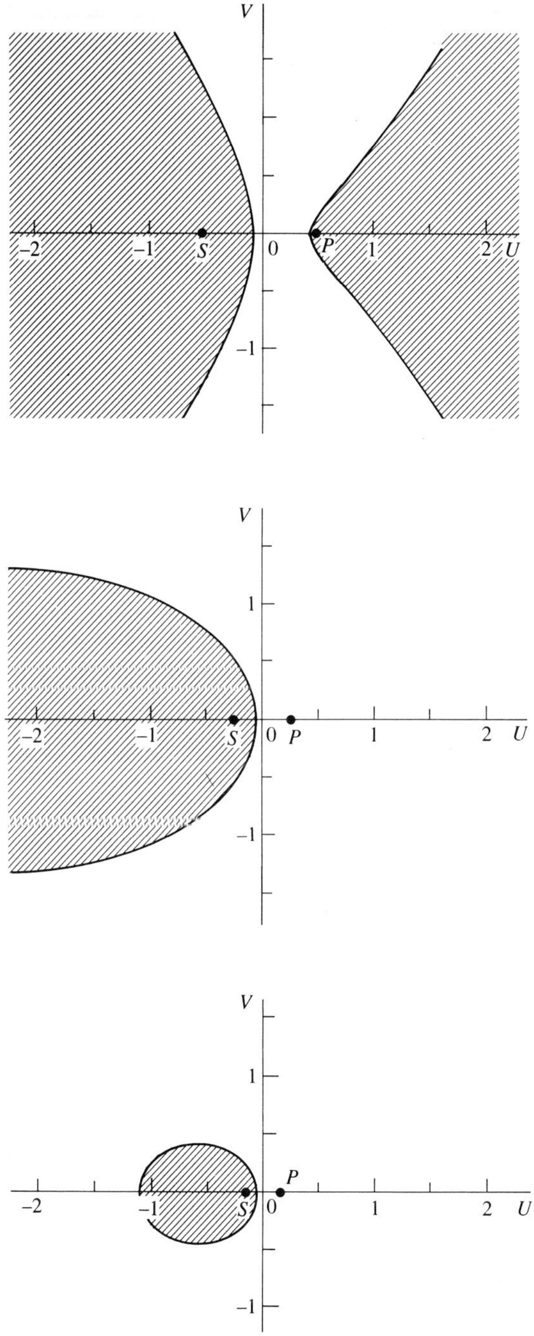

(a) d=100 cm

(b) d=50 cm

(c) d=30 cm

Fig. 3.14. Zones of cancellation and reinforcement (f = 100 Hz)

where, the shaded areas correspond to reinforcement and the plain areas correspond to cancellation. The axes are graduated in metres. The corresponding three-dimensional sketch of these is obtained by revolving each diagram around the U-axis.

It follows from Examples 3.4 and 3.5 that for a given tone of the component waves a decrease in the separation between the sources leads to an increase in the extent of zones of cancellation and a decrease in the extent of zones of reinforcement. Similarly, for a given separation between the sources the extent of zones of cancellation will increase and simultaneously the extent of zones of reinforcement will decrease with decreasing frequencies. Furthermore, the pattern of cancellation and reinforcement is found to be symmetric about a straight line joining locations of the primary and secondary sources. The latter is also evident from equation (3.67) which is equivalent to seeking loci of constant α–constant β.

3.5 Summary

A parametric analysis of the process of field cancellation by active means in a three-dimensional non-dispersive acoustic medium is given. This involves the derivation of the field cancellation factor, which gives a quantitative measure of cancellation, in terms of amplitude and phase parameters of the primary and secondary waves at an observation point in the medium, and the necessary conditions of cancellation on which the design of an ANC system can be based.

There are basically two sets of conditions involved in the process of cancellation: (a) source-related and (b) geometry-related. An artificially generated secondary source of noise is required to produce a wave that is coherent with the unwanted noise over the existing frequency range of the primary wave. This, in turn, will require the electronic components such as sensors, amplifiers and loudspeakers to have suitable characteristics so that an undistorted stationary secondary wave is generated.

In propagating through the medium the primary and secondary waves are affected in amplitude and in phase by the properties of the medium. This results in a combined field forming patterns of cancellation and reinforcement. A consideration of the functional relation between field cancellation factor and geometry-related parameters, such as the separation between the sources and the location of an observation point in the medium, leads to the three-dimensional description of cancellation. It is found from such a consideration that it is possible to cancel the noise over large volumes of the medium by proper adjustment of the separation between the sources and, furthermore, with a carefully chosen secondary source, even obtain an overall elimination of the noise.

3.6 References

Carlson, A.B. (1975). *Communication systems*, (2nd edn). McGraw-Hill, London.

Leitch, R.R. and Tokhi, M.O. (1987), Active noise control systems. *IEE Proceedings*, Part A, **134**, (6), 525–46.

Leventhal, H.G. (1976). Developments in active attenuators. *Proceedings of Noise Control Conference*, Warsaw, pp. 33–42.

Stanley, W.D. (1982). *Electronic communications systems*, Reston Publishing Company, Virginia.

Vogt, M. (1976). General conditions of phase cancellation in an acoustic field. *Archives of Acoustics*, **1**, (2), 109–25.

4

CONTROLLER DESIGN

4.1 Introduction

It follows from the theoretical consideration of field cancellation that, from a practical point of view, active attenuation (or cancellation) of noise through wave interference consists of the three main processes of detection, negation, and superposition.

The aim in the process of detection is to obtain a signal coherent with the unwanted noise over the frequency range of interest. Thus, it is important to use a detector with suitable response characteristics to provide such information. Microphones are commonly used as detectors. These are found to have reasonably flat amplitude and nearly linear phase characteristics. Here the detector is placed at a fixed distance relative to the primary source of noise giving a direct measure of the acoustic signal.

An alternative to direct detection is to measure a non-acoustic signal that is coherent with the noise (Dines 1982; Kim *et al.* 1984; Ross 1982, 1986). Many sources of noise, for instance, vibrate in coherence with the acoustic waves they emit. Thus, in these circumstances a measure of vibration can provide (indirectly) the required information on the noise. Flame noise is another example where indirect detection can be used. It has been reported that the coherence between light and noise can be as much as 99% (Dines 1982; Ross 1986). An advantage of indirect detection is that the acoustic feedback, due to secondary source radiation entering the detector, which can sometimes lead to instability, is avoided. However, indirect detection must be used with care to ensure that a good measure of the noise is obtained.

The most important part, and the main body, of ANC is the process of negation. The device performing this task (hereafter referred to as the controller) should be capable of not only adjusting the amplitude but also shifting the phase of each frequency component of the detected signal, accordingly, by 180°. Therefore, the controller is defined by a continuous transfer function representing the required amplitude and phase characteristics. Thus, when the processed wave emitted by the secondary source is superimposed on the primary wave destructive interference results at the observation point. The required controller characteristics are derived in this chapter. These are found to be dependent upon the characteristics of the transmission paths from the primary and secondary sources to the detection and observation points as well as the characteristics of the secondary source and transducers used in the system.

After the detected signal has been processed by the controller, it is then used to drive the secondary source. This leads to an acoustic wave that interacts with the noise from the primary source. Here a loudspeaker placed at a set distance from the primary source is used as a secondary source. The result of superimposing the acoustic waves can be observed by using an observer microphone. It follows from the above that the processes of detection and superposition are relatively straightforward, whereas the main task in ANC is the design of a suitable controller.

This chapter presents an analysis of the basic control structures common in ANC systems; namely, feedback control and feedforward control. Conditions for the robust performance of such systems are determined. A controller design procedure based on the optimum performance of the system is given. The resulting controller design equations are then related to the system geometry leading to an interpretation of the above conditions as constraints on the geometric composition of the system.

4.2 Active noise control structures

Active noise control requires a suitable interconnection of detector, controller, and secondary source. The various arrangements of detector(s), controller, and secondary source(s) in single/multiple detector and/or source configurations have resulted in various control structures for ANC systems. These structures can be classified into two basic types: feedback control and feedforward control. These structures are considered in this section and the controller design relations derived.

4.2.1 Feedback control structure

The geometric arrangement of the feedback control structure (FBCS) is shown schematically in Fig. 4.1. The primary source emits a wave $p(t)$. This is detected by a transducer (detector), placed at a fixed distance relative to the primary source, and transferred to a controller C. The controller is required to adjust the phase as well as the amplitude of each frequency component contained in the detected signal so that, when emitted by the secondary source, it results in zero sound pressure level at the detector location. Hence, the detector can at the same time be thought of as an observer.

The FBCS corresponds to that proposed by Lueg (1936) in his patent (Fig. 2.2(b)) where the controller is realized by an electronic transmission line providing a constant time delay and hence phase inversion of a particular frequency. Similarly, Olson used this structure in his electronic sound absorber with an amplifier as a controller (Olson 1956, 1961; Olson and May 1953). Following Olson's work a significant amount of consideration has been given to the investigation and performance improvement of this

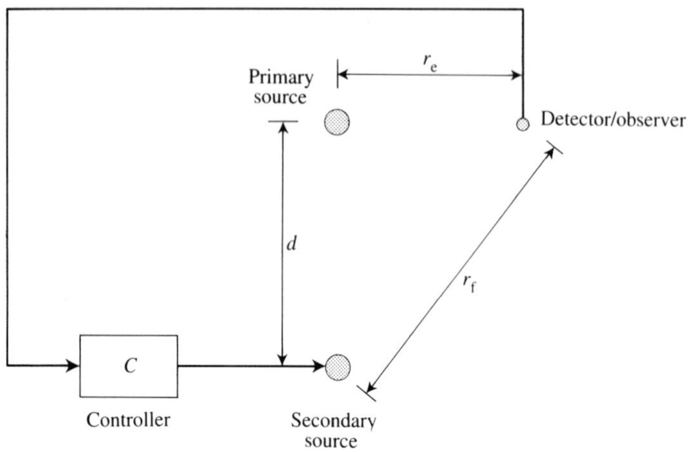

Fig. 4.1. Schematic diagram of FBCS

structure in various applications (Bleazey 1962; Chaplin *et al.* 1985; Short 1980; Trinder and Nelson 1983*a,b*).

Fig. 4.2 shows an equivalent block diagram of the FBCS in the complex frequency s domain, where:

$E(s)$ = transfer function of acoustic path between primary source and detector,
$F(s)$ = transfer function of acoustic path between secondary source and detector,
$M(s)$ = transfer function of detector,
$C(s)$ = transfer function of controller,
$N(s)$ = transfer function of necessary electronics, in addition to the controller, and
$L(s)$ = transfer function of secondary source.

The feedback nature of this structure is evident from Fig. 4.2. A (control) signal $S(s)$ from the secondary source is fed back to the detector, through the acoustic transmission path between the secondary source and detector. There, it adds to a (reference) signal $R(s)$ (thus giving positive feedback) to give rise to a (residual or error) signal $D(s)$. The detector signal $D(s)$ can be written as

$$D(s) = R(s) + F(s)S(s)$$

or

$$D(s) = E(s)P(s) + F(s)M(s)C(s)N(s)L(s)D(s).$$

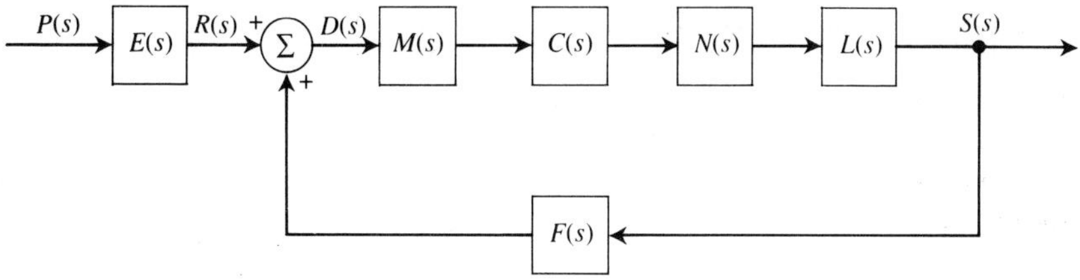

Fig. 4.2. Block diagram of the FBCS

Simplifying this yields the transfer function between the detector signal $D(s)$ and the primary source signal $P(s)$ as

$$\frac{D(s)}{P(s)} = \frac{E(s)}{1 - M(s)C(s)N(s)L(s)F(s)}. \tag{4.1}$$

Assuming the controller always reverses the polarity of the signal, equation (4.1) can thus be written as

$$\frac{D(s)}{P(s)} = \frac{E(s)}{1 + M(s)C'(s)N(s)L(s)F(s)} \tag{4.2}$$

where $C'(s) = -C(s)$. Equation (4.2) corresponds to the standard negative feedback structure.

The objective in an FBCS is to reduce the detected signal to zero. Thus,

$$D(s) = 0. \tag{4.3}$$

This, using equation (4.2) and the Shwartz inequality (Tropper 1969), means that the following should hold

$$|D(s)| \leq |1 + M(s)C'(s)N(s)L(s)F(s)|^{-1}\,|E(s)|\,|P(s)|. \tag{4.4}$$

Equation (4.4) implies that for a given detector, necessary electronics, and secondary source the factor $|1 + M(s)C'(s)N(s)L(s)F(s)|^{-1}$ should tend to zero which in turn implies that $C'(s)$ or, equivalently, the controller transfer function $C(s)$ should have infinite amplitude:

$$|C(s)| = \infty. \tag{4.5}$$

In practice this is not feasible as the frequency characteristics $M(s)$, $N(s)$, $L(s)$, $F(s)$, and $C(s)$ itself will induce instability in the loop. The maximum value of $|C(s)|$ can be determined from the Nyquist stability criterion, and for satisfactory operation a suitable stability margin is required. This represents the major limitation of the feedback control structure: to achieve good cancellation a high-gain controller is required but this can produce instability in the system. Therefore, in designing the controller the loop formed by $M(s)$, $C(s)$, $N(s)$, $L(s)$, $F(s)$ should be analysed and a compromise made between controller gain and system performance.

4.2.2 Feedforward control structure

A schematic diagram of the geometric arrangement of the feedforward control structure (FFCS) is shown in Fig. 4.3. The primary source emits a wave $p(t)$. This is detected by a sensor placed at a distance r_e relative to the primary source and a distance r_f relative to the secondary source and fed to the controller C. After the detected signal has been adjusted in phase and amplitude it is emitted by the secondary source to be superimposed on the unwanted noise. The result of this superposition is observed at an observation point located at a distance r_g relative to the primary source and a distance r_h relative to the secondary source.

The FFCS corresponds to that proposed by Lueg (1936) in his patent (Fig. 2.2(d)) where the time delay is implemented by the physical separation between the primary and secondary sources. A significant amount of consideration has subsequently been given to this structure in various ANC applications. Conover (1956, 1957), Hesselman (1978), and Ross (1978) have employed this structure in the cancellation of transformer noise. Ross (1981) has considered the design, Roure (1985) has analysed the stability, and Eriksson and his associates have considered the implementation of this structure in the cancellation of one-dimensional duct noise (Eriksson and Allie 1987; Eriksson *et al.* 1987, 1988*a,b*; Munjal and Eriksson 1988). Nelson and his co-workers have analysed the performance of this structure in the cancellation of enclosed sound fields (Bullmore *et al.* 1987; Elliott and Nelson 1987; Elliott *et al.* 1987; Nelson *et al.* 1987*a,b*). Tokhi and Leitch have considered the design, performance, and implementation of this structure in

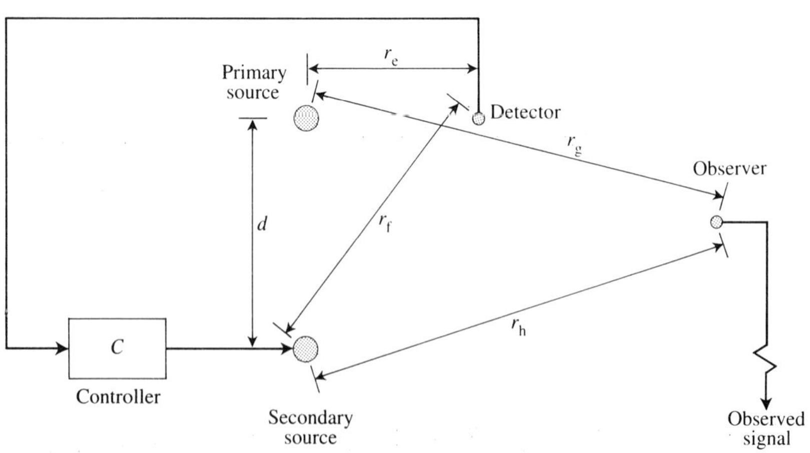

Fig. 4.3. Schematic diagram of FFCS

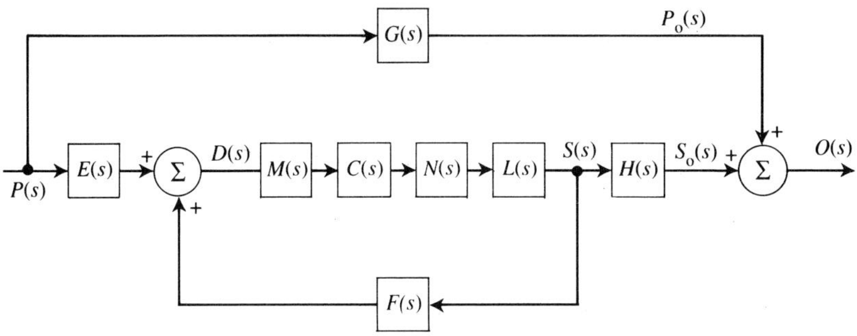

Fig. 4.4. Block diagram of the FFCS

the cancellation of noise in a three-dimensional propagation medium (Leitch and Tokhi 1986, 1987; Tokhi and Leitch 1989).

A block diagram of Fig. 4.3 is shown in the complex frequency s domain in Fig. 4.4 where:

$E(s)$ = transfer function of path r_e,
$F(s)$ = transfer function of path r_f,
$G(s)$ = transfer function of path r_g,
$H(s)$ = transfer function of path r_h,
$M(s)$ = transfer function of the detector,
$C(s)$ = transfer function of the controller,
$N(s)$ = transfer function of necessary electronics, and
$L(s)$ = transfer function of the secondary source.

From the block diagram in Fig. 4.4 the detector output $D(s)$ and secondary source output $S(s)$ are

$$D(s) = E(s)P(s) + F(s)S(s) \tag{4.6}$$

and

$$S(s) = M(s)C(s)N(s)L(s)D(s). \tag{4.7}$$

Substituting for $D(s)$ from equation (4.6) into equation (4.7) yields

$$S(s) = M(s)C(s)N(s)L(s)[E(s)P(s) + F(s)S(s)].$$

Simplifying this yields the transfer function $A(s)$ between the secondary source and primary source outputs as

$$A(s) = \frac{S(s)}{P(s)} = \frac{M(s)C(s)N(s)E(s)}{1 - M(s)C(s)N(s)L(s)F(s)}. \tag{4.8}$$

Thus, using $A(s)$, the block diagram in Fig. 4.4 can be represented as Fig. 4.5 (compare this diagram with Fig. 3.2).

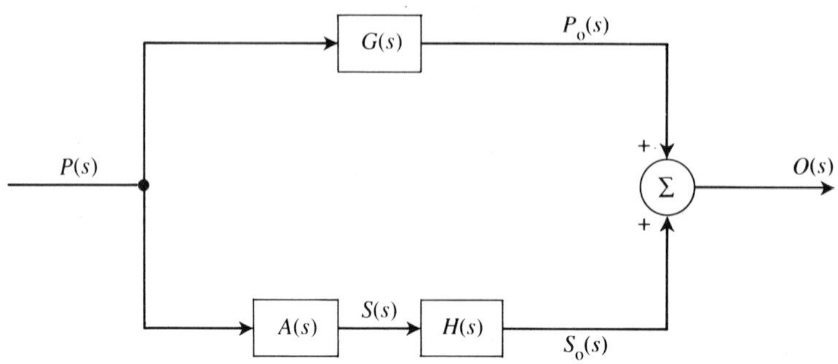

Fig. 4.5. Simplified block diagram of the FFCS

It follows from Figs 4.4 and 4.5 that, in the FFCS, the secondary path through $C(s)$ to the observation point attempts to compensate for the primary path through $G(s)$ to the observation point in such a way that the superposition of the primary and secondary waves results in cancellation at the observation point.

As seen in Fig. 4.3 the detector gives a combined measure of the primary and secondary waves reaching the detection point, through the acoustic paths r_e and r_f, respectively. The path through r_f between the detector and secondary source thus forms an acoustic feedback path. The closed loop thus formed can cause the system to become unstable, therefore, a careful consideration of this loop will be necessary in a design stage. This is considered in Chapter 5.

The objective with the FFCS is to reduce the level of noise to zero at the observation point. This requires that the observed primary and secondary signals should be equal in amplitude and opposite in phase:

$$P_o(s) = -S_o(s). \tag{4.9}$$

From Fig. 4.4 the primary and secondary signals $P_o(s)$ and $S_o(s)$ are

$$\begin{aligned} P_o(s) &= G(s)P(s) \\ S_o(s) &= H(s)S(s). \end{aligned} \tag{4.10}$$

The secondary source signal $S(s)$ follows from Fig. 4.5:

$$S(s) = A(s)P(s). \tag{4.11}$$

Substituting for $P_o(s)$ and $S_o(s)$ from equations (4.10) into equation (4.9), using equation (4.11), and simplifying yields

$$G(s) = -H(s)A(s). \tag{4.12}$$

Equation (4.12) is the required relation under which full cancellation of noise is achieved at the observation point. Substituting for $A(s)$ from equation (4.8) into equation (4.12) and solving the result for $C(s)$ yields

$$C(s) = \frac{G(s)}{M(s)N(s)L(s)\Delta(s)} \tag{4.13}$$

where

$$\Delta(s) = F(s)G(s) - E(s)H(s). \tag{4.14}$$

Equation (4.13) is the required controller transfer function for optimum cancellation of noise over the frequency range of interest.

Note in Fig. 4.3 that if the observation point is moved to coincide with the detection point then the FBCS results. Therefore, subsequent developments are mainly considered in terms of FFCS, whereas those in terms of the FBCS are considered as special cases.

4.3 Geometry-related constraints

It follows from equation (4.13) that, for a particular detector and secondary source with necessary electronic components, the controller characteristics required for optimum cancellation of noise are dependent on the characteristics of the acoustic paths from the primary and secondary sources to the detection and observation points. A change in the location of the detector and/or observer with respect to the primary and secondary sources is reflected in the required controller transfer function. In particular, if the detector and observer are located such that $\Delta(s)$ in equation (4.13) becomes zero then the critical situation of an infinite-gain controller requirement arises. Such a situation is a constraint on the geometric composition of the system and is considered as a practical limitation in the design of the controller. The locus of detection and observation points in the medium leading to such a situation is therefore of crucial interest. This locus is derived and discussed in this section.

4.3.1 Locus of an infinite-gain controller

If $\Delta(s)$ in equation (4.13) becomes zero then infinite gain is required of the controller. Under such a situation equation (4.14) for $s = j\omega$ yields

$$F(j\omega)G(j\omega) = E(j\omega)H(j\omega)$$

or

$$\frac{F(j\omega)}{E(j\omega)} = \frac{H(j\omega)}{G(j\omega)}. \tag{4.15}$$

$E(j\omega)$, $F(j\omega)$, $G(j\omega)$, and $H(j\omega)$ are the transfer functions of the acoustic paths through the distances r_e, r_f, r_g, and r_h, respectively; thus

$$E(j\omega) = \frac{A}{r_e}\,e^{-j2\pi r_e/\lambda}; \qquad F(j\omega) = \frac{A}{r_f}\,e^{-j2\pi r_f/\lambda}$$

$$G(j\omega) = \frac{A}{r_g}\,e^{-j2\pi r_g/\lambda}; \qquad H(j\omega) = \frac{A}{r_h}\,e^{-j2\pi r_h/\lambda} \tag{4.16}$$

where λ is the wavelength, and A is a constant.

Substituting for $E(j\omega)$, $F(j\omega)$, $G(j\omega)$, and $H(j\omega)$ from equations (4.16) into equation (4.15) and simplifying yields

$$\left(\frac{r_e}{r_f}\right)e^{-j(r_f - r_e)2\pi/\lambda} = \left(\frac{r_g}{r_h}\right)e^{-j(r_h - r_g)2\pi/\lambda}. \tag{4.17}$$

This equation is true if and only if the amplitudes as well as the exponents (phases) on either side of the equation are equal. Equating the amplitudes and the phases accordingly yields

$$\frac{r_e}{r_f} = \frac{r_g}{r_h} = a$$

$$r_f - r_e = r_h - r_g \tag{4.18}$$

where a, the distance ratio, is a positive real number.

Equations (4.18) define the locus of points for which $\Delta(j\omega) = 0$, and thus the controller is required to have infinitely large gain. Note that these equations are in terms of the distances r_e, r_f, r_g, and r_h only. Therefore, the critical situation of $\Delta(j\omega) = 0$ is determined only by the locations of the detector and observer relative to the primary and secondary sources.

Eliminating r_f and r_h in equations (4.18) yields

$$r_e - \frac{r_e}{a} = r_g - \frac{r_g}{a}$$

or

$$r_e(a - 1) = r_g(a - 1). \tag{4.19}$$

The two possible situations, namely, $a = 1$ and $a \neq 1$, are considered separately.

Unity distance ratio For a unity distance ratio $a = 1$, equation (4.19) yields the identity $0 = 0$. Therefore, substituting for $a = 1$ into equations (4.18) yields the locus of points for which infinite gain is required of the controller as

$$\frac{r_e}{r_f} = 1 \quad \text{and} \quad \frac{r_g}{r_h} = 1. \tag{4.20}$$

If the locations of the primary and secondary sources are fixed then each of equations (4.20) defines a plane surface perpendicularly bisecting the line joining the primary and secondary source locations (see Appendix A). This plane for the primary and secondary sources, located respectively, at points (0, 0, 0) and (u_s, v_s, w_s) with a distance d apart in a three-dimensional UVW-space (see Fig. 4.6) is given by

$$\frac{u}{\left(\dfrac{d^2}{2u_s}\right)} + \frac{v}{\left(\dfrac{d^2}{2v_s}\right)} + \frac{w}{\left(\dfrac{d^2}{2w_s}\right)} = 1 \tag{4.21}$$

which intersects the U-, V-, and W-axes, respectively, at points $(d^2/2u_s, 0, 0)$, $(0, d^2/2v_s, 0)$, and $(0, 0, d^2/2w_s)$. This plane is shown in Fig. 4.7. If the detector is placed at any point on this plane and if at the same time the observer location coincides with a point on this plane then the 'critical situation' of equation (4.15) occurs and for optimum cancellation of noise the controller is required to have infinitely large gain.

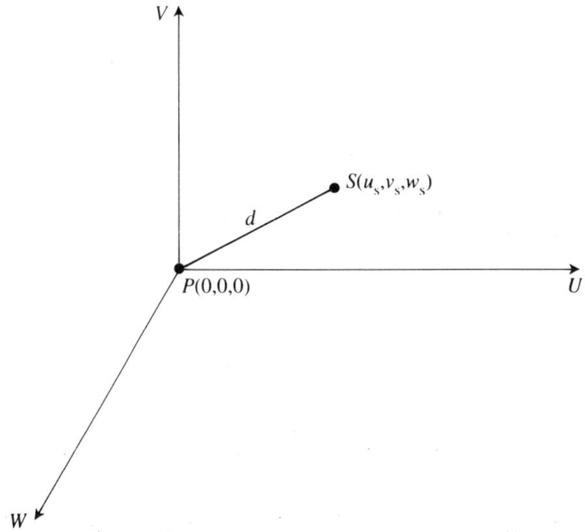

Fig. 4.6. Location of primary and secondary sources in three-dimensional coordinates

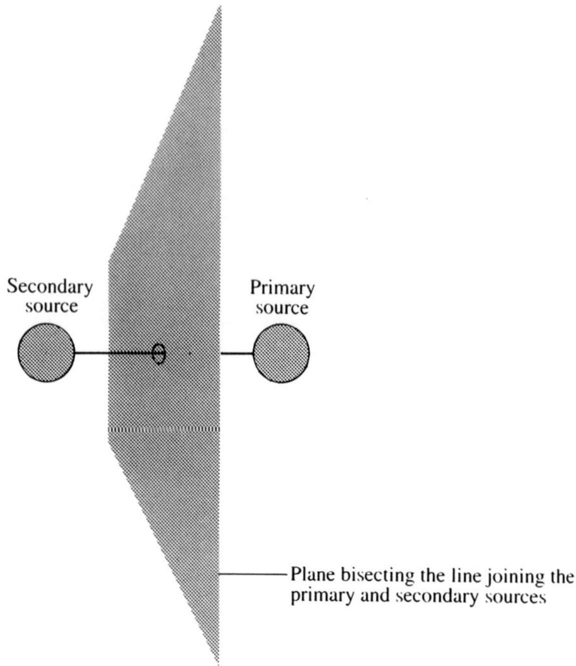

Fig. 4.7. Plane of infinite-gain controller

Non-unity distance ratio For a non-unity distance ratio $a \neq 1$, equations (4.18) and (4.19) yield the locus of points for which an infinitely large gain is required of the controller as

$$\frac{r_e}{r_f} = a, \qquad \frac{r_g}{r_h} = a \qquad \text{and} \qquad \frac{r_e}{r_g} = 1. \tag{4.22}$$

It follows from Appendix A that the first two relations in equation (4.22) define a spherical surface. These surfaces for the primary and secondary sources located as in Fig. 4.6 are both defined by

$$\left[u + \frac{a^2 u_s}{1 - a^2}\right]^2 + \left[v + \frac{a^2 v_s}{1 - a^2}\right]^2 + \left[w + \frac{a^2 w_s}{1 - a^2}\right]^2 = \left[\frac{ad}{1 - a^2}\right]^2 \tag{4.23}$$

which has a radius $R = ad/|1 - a^2|$ and centre located along the line PS, joining the primary and secondary source locations, at the point $Q(-a^2 u_s/(1 - a^2),\ -a^2 v_s/(1 - a^2),\ -a^2 w_s/(1 - a^2))$. This is shown in two dimensions in Fig. 4.8.

The third relation in equation (4.22) requires the equality of the distances between the detector and primary source r_e and between the observer and

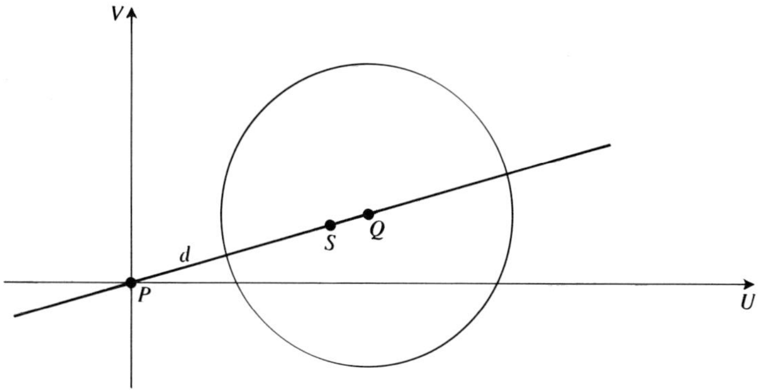

Fig. 4.8. Locus of constant distance ratio

primary source r_g. The locus of such points in the three-dimensional UVW-space of Fig. 4.6 (for, say, constant r_e) is a sphere with centre at the primary source location and radius equal to r_e:

$$u^2 + v^2 + w^2 = r_e^2. \tag{4.24}$$

This is shown in two dimensions in Fig. 4.9. Therefore, the locus of points defined by equations (4.22) is given by the intersection of the two spheres in equations (4.23) and (4.24). Such a locus is a circle, hereafter referred to as the infinite-gain controller (IGC) circle, in a plane at right angles to the line joining the centres of the spheres. The centre of the circle is the point of intersection of the plane and the line. Since the centres of the two spheres are located along the line PS joining the primary and secondary source locations (Fig. 4.6), the centre of the IGC circle is thus the point of intersection of the plane and the line PS.

To investigate the variation of the IGC circle in terms of its radius and the location of its centre in the three-dimensional UVW-space of Fig. 4.6, let the detector be located at point D with coordinates (u_d, v_d, w_d) and distances r_e and r_f relative to the primary and secondary sources, respectively, as shown in Fig. 4.10. Substituting for $u^2 + v^2 + w^2$ from equation (4.24) into equation (4.23) and simplifying yields the plane of the IGC circle as

$$\frac{u}{\left(\dfrac{B}{u_s}\right)} + \frac{v}{\left(\dfrac{B}{v_s}\right)} + \frac{w}{\left(\dfrac{B}{w_s}\right)} = 1 \tag{4.25}$$

where

$$B = \frac{1}{2}\left[d^2 - \left(\frac{1}{a^2} - 1\right)r_e^2\right] = \frac{1}{2}\left[d^2 - (r_f^2 - r_e^2)\right]. \tag{4.26}$$

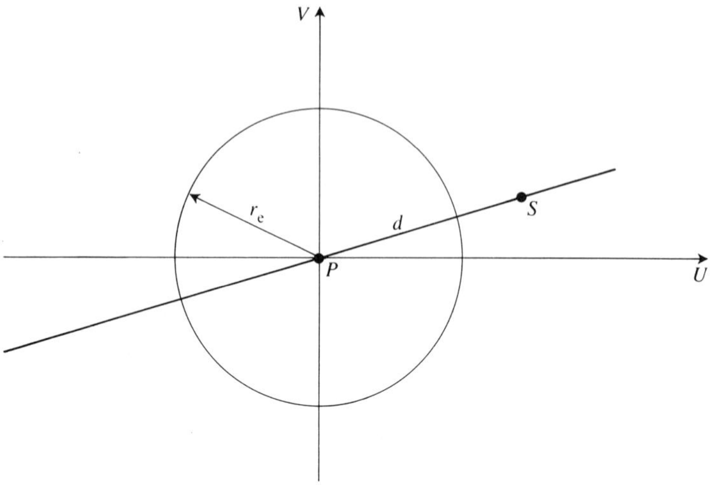

Fig. 4.9. Locus of points $r_e = r_g$

Equation (4.25) defines a plane surface on which the IGC circle is residing. It is found through the procedure described in Appendix A that the line *PS* passing through the primary and secondary source locations is at right angles

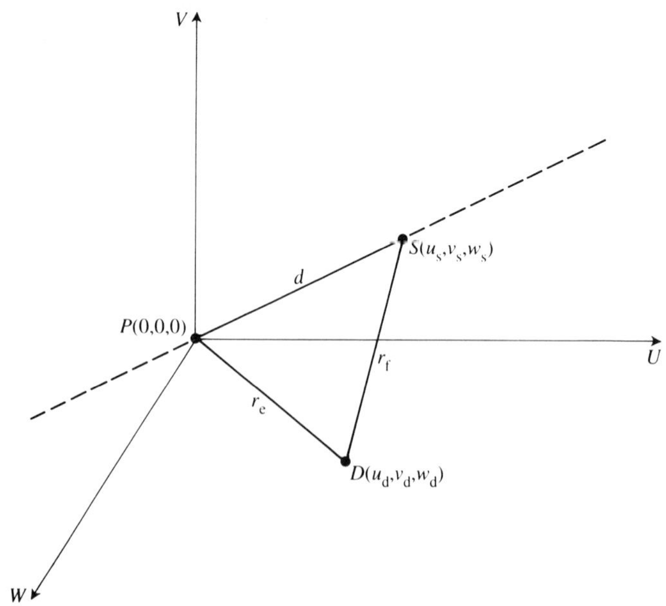

Fig. 4.10 Location of detector relative to primary and secondary sources

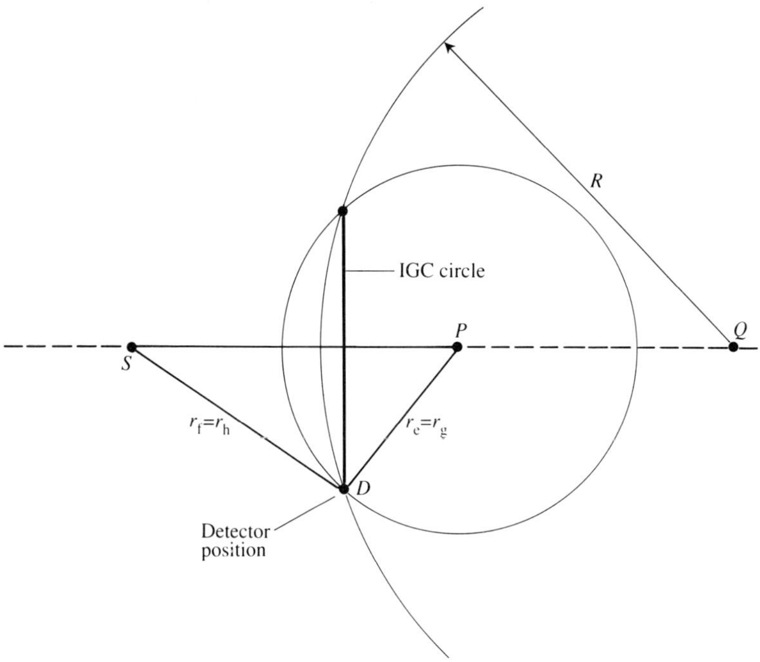

Fig. 4.11. Formation of the IGC circle

to this plane. This is shown in two dimensions in Fig. 4.11. The corresponding IGC circle is shown in Fig. 4.12 where r_c is the radius of the IGC circle.

The quantity B in equation (4.26) gives a measure of the intersection of the plane in equation (4.25) with the coordinate axes and, thereby, with the line PS passing through the primary and secondary source locations. It is evident from equation (4.26) that B is dependent on d, r_e, and r_f or, for constant d, it depends on the location of the detector only. If θ represents the angle formed by lines PD and PS in a plane on which both these lines reside (see Fig. 4.13) then the following holds

$$r_f^2 = d^2 + r_e^2 - 2r_e d \cos\theta. \tag{4.27}$$

Substituting for r_f^2 from equation (4.27) into equation (4.26) yields

$$B = r_e d \cos\theta. \tag{4.28}$$

Therefore, as the location of the detection point changes in the medium, the limits for B are found to be

$$|B| \le r_e d. \tag{4.29}$$

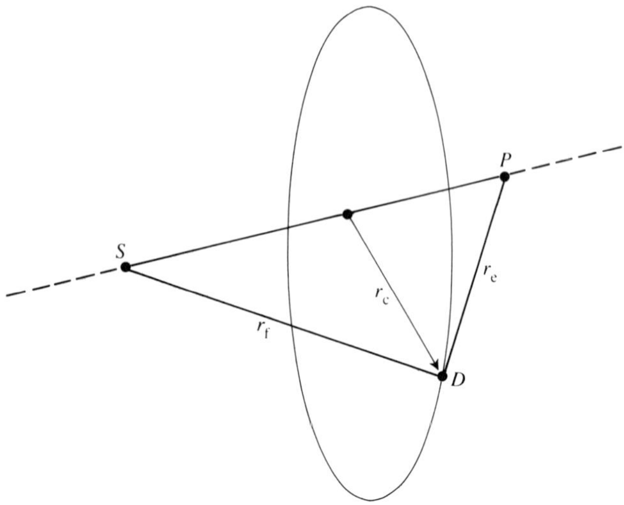

Fig. 4.12. The IGC circle

This variation in relation to the location of the plane of the IGC circle is shown in two dimensions in Fig. 4.13. For $r_f^2 - r_e^2 > d^2$, in which case $0 > B > -r_e d$, the plane passes through points along the line PS which are outside the range (P, S) and their distances from P are smaller than those from point S. As $r_f^2 - r_e^2$ decreases the plane moves towards point P. At point P, where $r_f^2 - r_e^2 = d^2$, the quantity B is zero. If $r_f^2 - r_e^2$ is further decreased so that a is still greater than unity the plane moves towards the midpoint between P and S. Midway between points P and S the distance ratio a is unity ($r_f = r_e$), $B = \frac{1}{2}d^2$ and the plane coincides with the plane in equation (4.21); see Fig. 4.7. After the point midway between P and S ($0 < a < 1$), as the distance ratio a is decreased, the plane moves towards point S and further beyond it. At point S, where $r_e^2 - r_f^2 = d^2$, the quantity B is equal to d^2.

From Fig. 4.13 the radius r_c of the IGC circle can be written as

$$r_c = r_e \sin\theta, \qquad 0 \leq \theta \leq \pi. \tag{4.30}$$

Thus, the radius of the IGC circle is dependent on the distance r_e between the primary source and the detector and the sine of the angle formed by the line joining the primary source and detector with the line joining the primary and secondary sources. The maximum value of the radius, $r_{c\,max}$, is r_e and occurs at the situation where the plane of the IGC circle intersects the line PS at point P (see Fig. 4.13);

$$r_{c\,max} = r_e. \tag{4.31}$$

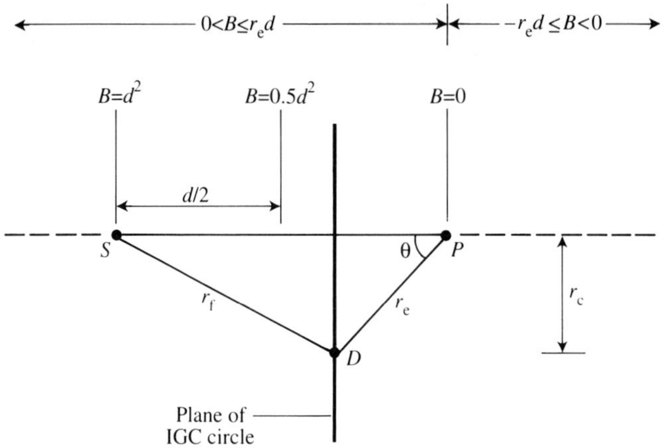

Fig. 4.13 Relative location of the IGC circle

Movement of the plane to either side of point *P* will result in a decrease in the radius. At the extreme case where the line *PD* is in alignment with the line *PS* (θ is either 0° or 180°) the radius r_c is zero. In general, for constant values of the angle θ the radius r_c is directly proportional to the distance r_e between the primary source and the detector. This implies that in order for r_c to be minimized the detector is required to be placed as close to the primary source as possible.

It follows from the above that the requirement of an infinitely large gain controller in an FFCS is directly linked with the locations of the detector and observer relative to the primary and secondary sources. This derives from the dependence of the controller characteristics on the transfer characteristics of the acoustic paths from the detector and observer to the primary and secondary sources, which demand a particular controller transfer function for a particular combination of the detector and observer locations in the medium. The above analysis reveals that combinations of the detector and observer locations in the medium exist, which, for optimum cancellation of noise, require the controller to have infinitely large gain. These locations form the locus of infinite-gain controller requirement which are:

1. If the detector and observer are equidistant from the sources the locus is a plane surface that perpendicularly bisects the line joining the locations of the primary and secondary sources.
2. If the detector and observer are not equidistant from the sources the locus is a circle, with centre along the line joining the locations of the primary and secondary sources, and on a plane that is parallel with the plane in 1. The radius of the circle is given by the shortest distance between the detector and the line.

Note in equation (4.22) that if the first two of these relations are divided side by side (assuming $a \neq 0$) then the following equivalent equations are obtained

$$\frac{r_e}{r_g} = 1 \qquad \text{and} \qquad \frac{r_f}{r_h} = 1. \tag{4.32}$$

This means that, starting with equations (4.32) rather than equation (4.22), will also lead to exactly the same results obtained in the preceding paragraphs.

The crucial situation of an infinite-gain controller requirement was shown in the above, on the basis of using omnidirectional sensors and sources, to exist with the general ANC structure. The problem can be avoided by confining the detector and observer to regions of the medium that are outside the loci in 1. and 2. above. A further possibility of avoiding this problem is to isolate the detector from secondary source radiation. A method of achieving this is to use a unidirectional detector such that it is subjected to the primary wave only. Alternatively, indirect detection, as discussed earlier, can be used. Either of these methods, in terms of the FFCS, are equivalent to making $F(s) = 0$, under which the controller transfer function in equation (4.13) becomes

$$C(s) = -\frac{G(s)}{\mathrm{M}(s)\mathrm{N}(s)\mathrm{L}(s)\mathrm{E}(s)\mathrm{H}(s)} \qquad \text{for } F(s) = 0. \tag{4.33}$$

Roure (1985) has discussed the use of unidirectional microphones as an attempt to avoid the infinite-gain controller requirement in the case of one-dimensional propagation (duct noise). Others have reported alternative configurations of multiple-detector, multiple-source for the one-dimensional duct noise problem as attempts at isolating the detector from secondary source radiation (Jessel and Mangiante 1972; Leventhal and Eghtesadi 1979; Swinbanks 1973).

4.3.2 Feedback control structure

As noted earlier, if the observer in an FFCS shown in Fig. 4.3 moves to coincide with the detector then the FBCS is obtained. In such a process the distances r_g and r_h are effectively made to approach the distances r_e and r_f, respectively. This, in terms of the transfer functions $E(s)$, F(s), G(s), and $H(s)$, leads to

$$\begin{aligned} G(s) &= E(s) \\ H(s) &= F(s). \end{aligned} \tag{4.34}$$

Projecting the above modifications into the controller design equations of an FFCS the corresponding controller design equations for the FBCS are obtained. Thus, substituting for $G(s)$ and $H(s)$ from equations (4.34) into equation (4.14) yields

$$\Delta(s) = E(s)F(s) - E(s)F(s)$$

or

$$\Delta(s) = 0. \tag{4.35}$$

Equation (4.35) corresponds to the critical situation of the infinite-gain controller requirement discussed above. Therefore, as also shown earlier from a different perspective, for optimum cancellation of noise the FBCS will always require a controller with infinitely large gain. With a practically acceptable compromise between system performance and controller gain, and careful consideration of the stability of the system, reasonable amounts of cancellation of noise can be achieved with this structure.

4.4 Summary

An analysis of the basic control structures and controller design procedure for ANC systems in three-dimensional propagation is presented.

The optimum cancellation of noise requires a controller with a frequency-dependent transfer function that can produce a wave which is an exact mirror image of the noise so that, when superimposed on the noise, it results in silence. The characteristics of such a controller are found to be dependent upon the transfer characteristics of transducers, secondary source, and propagation paths from the primary and secondary sources to both the detection and observation points.

The dependence of the controller characteristics on the characteristics of system components and geometry can sometimes lead to practical difficulties in the controller design and system stability. A particular combination of these characteristics requires a controller of a particular transfer function. A change in any of these characteristics, such as location of the detector and/or observer, requires a controller with a new transfer function to suit the new situation. In particular there are combinations of detector and observer locations which lead to the critical situation of an infinite-gain controller requirement. Two situations in general lead to the IGC requirement.

1. When both the observer and detector are equidistant from the primary and secondary sources. In an FBCS, where both the detection and observation points are the same, this situation corresponds to the detector being on a plane that perpendicularly bisects the line joining the locations

of the sources. In an FFCS, this situation corresponds to when both the detector and observer are on this plane.

2. When the ratio of the distances from the detector and the observer to the primary source and the ratio of the distances from the detector and the observer to the secondary source are each equal to unity. This arises when the detector and the observer are both on a circle (the IGC circle) which is in a plane that makes a 90° angle with the line passing through the locations of the primary and secondary sources. In an FBCS, where the detector and observer are both located at the same point, this situation of an IGC requirement is always satisfied. In an FFCS, however, it is possible to minimize the region of space occupied by the IGC circle by a proper geometrical arrangement of system components.

4.5 References

BLEAZEY, J.C. (1962). Electronic sound absorber. *Journal of Audio Engineering Society*, **10**, (2), 135–9.

BULLMORE, A.J., NELSON, P.A., CURTIS, A.R.D., and ELLIOTT, S.J. (1987). The active minimization of harmonic enclosed sound fields, Part II: A computer simulation. *Journal of Sound and Vibration*, **117**, (1), 15–33.

CHAPLIN, G.B.B., JONES, A., and JONES, O. (1985). *Methods and apparatus for low frequency active attenuation*. US Patent No. 4 527 282.

CONOVER, W.B. (1956). Fighting noise with noise. *Noise Control*, **92**, 78–82, 92.

CONOVER, W.B. (1957). *Noise reducing system for transformers*. US Patent No. 2 776 020.

DINES, P.J. (1982). Comparison of least squares estimation and impulse response techniques for active control of flame noise. *Proceedings of the Institute of Acoustics*, November, F2.1–F2.4.

ELLIOTT, S.J. and NELSON, P.A. (1987). The active control of enclosed sound fields. *Proceedings of National Conference on Noise Control Engineering*, Pennsylvania, pp. 359–64.

ELLIOTT, S.J., CURTIS, A.R.D., BULLMORE, A.J., and NELSON, P.A. (1987). The active minimization of harmonic enclosed sound fields, Part III: Experimental verification. *Journal of Sound and Vibration*, **117**, (1), 35–58.

ERIKSSON, L.J. and ALLIE, M.C. (1987). A digital sound control system for use in turbulant flows. *Proceedings of National Conference on Noise Control Engineering*, Pennsylvania, pp. 365–70.

ERIKSSON, L.J., ALLIE, M.C., and GREINER, R.A. (1987). The selection and application of an IIR adaptive filter for use in active sound attenuation. *IEEE Transactions on Acoustics, Speech, and Signal Processing*, **35**, (4) 433–7.

ERIKSSON, L.J., ALLIE, M.C., BREMIGAN, C.D., and GREINER, R.A. (1988*a*). Active noise control using adaptive digital signal processing. *Proceedings of the IEEE International Conference on Acoustics, Speech, and Signal Processing*, New York, **5**, 2594–7.

ERIKSSON, L.J., ALLIE, M.C., BREMIGAN, C.D., and GREINER, R.A. (1988*b*). Active noise control and specifications for fan noise problems. *Proceedings of Noise-Con 88: National Conference on Noise Control Engineering*, Indiana, pp. 273–8.

HESSELMAN, N. (1978). Investigation of noise reduction on a 100 KVA transformer tank by means of active methods. *Applied Acoustics*, **11**, (1), 27–34.

JESSEL, M. and MANGIANTE, G.A. (1972). Active sound absorbers in an air duct. *Journal of Sound and Vibration*, **23**, (3), 383–90.

KIM, C.D., MORIKAWA, S., ABE, M., and KIDO, K. (1984). Active suppression of noise in microphone line using vibration of noise source. *Proceedings of Inter-noise 84: International Conference on Noise Control Engineering*, Honolulu, **I**, 469–72.

LEITCH, R.R. and TOKHI, M.O. (1986). The implementation of active noise control systems using digital signal processing techniques. *Proceedings of the Institute of Acoustics*, **8**, Part 1, 149–57.

LEITCH, R.R. and TOKHI, M.O. (1987). Active noise control systems. *IEE Proceedings*, Part A, **134**, (6), 525–46.

LEVENTHAL, H.G. and EGHTESADI, Kh. (1979). Active attenuation of noise: monopole and dipole systems. *Proceedings of Inter-noise 79: International Conference on Noise Control Engineering*, Warsaw, **I**, 175–80.

LUEG, P. (1936). *Process of silencing sound oscillations.* US Patent No. 2 043 416.

MUNJAL, M.L. and ERIKSSON, L.J. (1988). An analytical, one-dimensional, standing-wave model of a linear active noise control system in a duct. *The Journal of the Acoustical Society of America*, **84**, (3), 1086–93.

NELSON, P.A., CURTIS, A.R.D., ELLIOTT, S.J., and BULLMORE. A.J. (1987*a*). The minimum power output of free field point sources and the active control of sound. *Journal of Sound and Vibration*, **116**, (3), 397–414.

NELSON, P.A., CURTIS, A.R.D., ELLIOTT, S.J., and BULLMORE, A.J. (1987*b*). The active minimization of harmonic enclosed sound fields, Part I: Theory. *Journal of Sound and Vibration*, **117**, (1), 1–13.

OLSON, H.F. (1956). Electronic control of noise, vibration and reverberation. *Journal of The Acoustical Society of America*, **28**, (5), 966–72.

OLSON, H.F. (1961). *Electronic sound absorber.* US Patent No. 2 983 790.

OLSON, H.F. and MAY, E.G. (1953). Electronic sound absorbers. *Journal of the Acoustical Society of America*, **25**, (6), 1130–6.

ROSS, C.F. (1978). Experiments on the active control of transformer noise. *Journal of Sound and Vibration*, **61**, (4), 473–80.

ROSS, C.F. (1981). Design of a broadband active sound control system. *Proceedings of the Institute of Acoustics*, February, 21–4.

ROSS, C.F. (1982). Quietening newspaper printing presses by active control. *Proceedings of the Institute of Acoustics*, November, F1.1–F1.4.

ROSS, C.F. (1986). Novel applications of active control techniques. *Proceedings of the Institute of Acoustics*, **8**, Part 1, 127–33.

ROURE, A. (1985). Self-adaptive broadband active sound control system. *Journal of Sound and Vibration*, **101**, (3), 429–41.

SHORT, W.R. (1980). Global low frequency active noise attenuation. *Proceedings of Inter-noise 80: International Conference on Noise Control Engineering*, Florida, **II**, 695–8.

SWINBANKS, M.A. (1973). Active control of sound propagation in long ducts. *Journal of Sound and Vibration*, **27**, (3), 411–36.

TOKHI, M.O. and LEITCH, R.R. (1989). Self-tuning active noise control. In *Adaptive filters*, IEE Colloquium, Digest No. 1989/46, 9/1–4.

TRINDER, M.C.J. and NELSON, P.A. (1983*a*). The acoustical virtual earth and its application to ducts with reflecting terminations. *Proceedings of Inter-noise 83: International Conference on Noise Control Engineering*, Edinburgh, **I**, 447–50.
TRINDER, M.C.J. and NELSON, P.A. (1983*b*). Active noise control in finite length ducts. *Journal of Sound and Vibration*, **89**, (1), 95–106.
TROPPER, A.M. (1969). *Linear algebra.* Nelson, London.

5

SYSTEM STABILITY

5.1 Introduction

It was noted in Chapter 4 that in an ANC system the detector gives a combined measure of the primary and secondary waves reaching the detection point. The secondary source radiation reaching the detector thus gives rise to acoustic feedback. Therefore, the closed loop formed by the detector, controller, secondary source, and the acoustic path between the secondary source and the detector (see Fig. 4.3) can cause the system to become unstable. A stability analysis of this loop is essential at a design stage.

Attempts have previously been made, in the case of one-dimensional propagation, to isolate the detector from the secondary source radiation (Jessel and Mangiante 1972; Leventhal and Eghtesadi 1979; Roure 1985; Swinbanks 1973). These primarily use multiple-detector multiple-source configurations or unidirectional sensors and sources so that the detector is subjected to the primary source radiation only, thereby avoiding acoustic feedback. A further possibility of avoiding feedback is to use indirect detection, by detecting the noise indirectly through measuring a non-acoustic signal that is coherent with the noise (Dines 1982; Kim *et al.* 1984; Ross 1982, 1986). The use of multiple-detector multiple-source configurations has obvious geometry-related limitations. The use of unidirectional sensors and sources rely on the assumption that the detector or source is unidirectional. As discussed in Chapter 3 the extent of a zone of cancellation is directly related to the separation between the primary and secondary sources. Therefore, if this separation is reduced to a minimum so as to increase the extent of cancellation zones then the above assumption as an attempt to avoid acoustic feedback may not be effective.

This chapter presents an analysis of the relative stability of the inherent feedback loop. For practical systems the use of absolute stability is not beneficial; a system having an extremely long oscillatory response is unlikely to be accepted and would be liable to instability under small parameter variations. In this respect a measure of the relative stability can provide a more acceptable design criterion. This procedure results in a robust design under stationary or slowly varying conditions. The frequency domain stability conditions are derived and interpreted as spatial conditions on the geometry of the ANC system.

5.2 The feedback loop

The loop formed by the detector, controller, secondary source, and the acoustic path between the secondary source and the detector is the only feedback loop present in the general ANC structure of Fig. 4.3. Therefore, from a stability point of view, it is sufficient to concentrate the analysis on this part of the system. Fig. 5.1 shows a block diagram of this loop, where $P_m(s)$ is the primary signal as measured at the detection point and $S(s)$ is the secondary source output signal. The controller transfer function $C(s)$ is assumed to be the required transfer function for optimum system performance described by equation (4.13).

From the block diagram in Fig. 5.1 the secondary signal $S(s)$ can be written as

$$S(s) = M(s)C(s)N(s)L(s)[P_m(s) + F(s)S(s)].$$

Simplifying this for the transfer function between $P_m(s)$ and $S(s)$ yields

$$\frac{S(s)}{P_m(s)} = \frac{M(s)C(s)N(s)L(s)}{1 + X(s)} \tag{5.1}$$

where

$$X(s) = -\, M(s)C(s)N(s)L(s(F(s). \tag{5.2}$$

Equation (5.1) has been written in a form to correspond to the standard negative feedback transfer function. It follows from this equation that, in order for the system in Fig. 5.1 to be stable, the denominator, $1 + X(s)$, should have roots in the left-hand side of the s-plane (D'Azzo and Houpis 1966). Note that equation (5.1) holds only if the primary source output is observable, that is, the detector is not located at a pressure null in the primary sound field.

5.3 The stability criterion

The stability analysis presented in this chapter is based on the Nyquist stability criterion which is a graphical method for determining the stability of

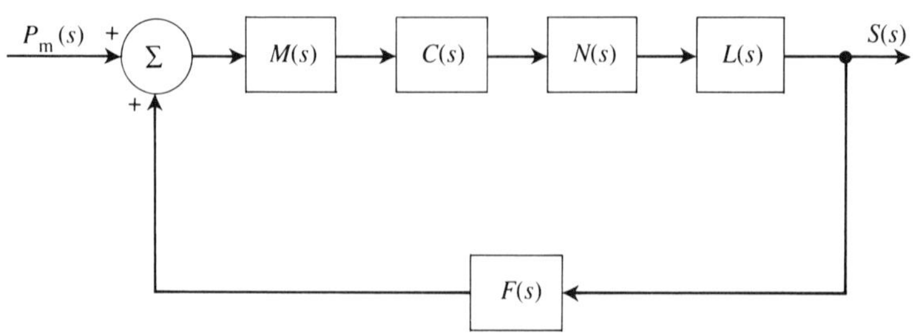

Fig. 5.1. The feedback loop

a system (D'Azzo and Houpis 1966). It is expressed in terms of the polar plot of the transfer function $X(s)$ for periodic waves ($s = j\omega$). Let $X(j\omega)$ have a magnitude $B(\omega)$ and a phase $\theta(\omega)$:

$$X(j\omega) = B(\omega)\, e^{j\theta(\omega)}. \tag{5.3}$$

Then, in accordance with the Nyquist stability criterion, for the system in Fig. 5.1, and hence the ANC system, to be stable the magnitude of $X(j\omega)$ at some frequency ω for which $\theta(\omega) = -(2n+1)\pi$; $n = 0, 1, 2, \ldots$, should be less than unity:

$$B(\omega) < 1 \qquad \text{when } \theta(\omega) = -(2n+1)\pi;\ n = 0, 1, 2, \ldots \tag{5.4}$$

where the negative angle indicates the (clockwise) direction of approach towards the π-axis on a polar plot of $X(j\omega)$. This can be expressed graphically by tracing the polar plot of $X(j\omega)$ from $\omega = 0$ to $\omega = \infty$ and observing each crossing of the π-axis, so that if the point -1 lies at the left-hand side then the system is considered to be stable, whereas if the point -1 lies at the right-hand side then the system is unstable.

5.3.1 Gain and phase margins

For practical systems a measure of absolute stability is not useful; a system that has an extremely long and oscillatory transient is unlikely to be accepted. In this respect a measure of the relative stability can provide a more acceptable design criterion. These, using frequency response plots, can be provided in terms of the gain and phase margins.

The gain margin is defined, at some frequency ω for which the phase $\theta(\omega)$ is $-180°$, as the additional gain k_g required to make the system unstable. This, in terms of the amplitude $B(\omega)$ of the transfer function $X(j\omega)$, is given by

$$k_g = \frac{1}{B(\omega)} \qquad \text{when } \theta(\omega) = -\pi. \tag{5.5}$$

Thus it follows from equations (5.4) and (5.5) that for a system to be stable the gain margin k_g must be greater than unity. A gain margin less than unity will mean that the system is unstable.

The phase margin is defined, at some frequency ω for which $B(\omega) = 1$, as the additional phase k_θ which makes the system unstable. This, in terms of the phase $\theta(\omega)$, is given by

$$K_\theta = \theta(\omega) + \pi \qquad \text{when } B(\omega) = 1. \tag{5.6}$$

Thus it follows from equations (5.4) and (5.6) that the phase margin k_θ is the amount of phase shift that would just produce instability. For minimum-phase systems (all zeros inside the stability region) to be stable the phase margin must be positive. A negative phase margin will mean that the system is unstable.

5.4 Spatial conditions

Let the controller in Fig. 5.1 have the transfer function required for optimum cancellation of noise, as given in equation (4.13). Substituting for $C(s)$ from equation (4.13) into equation (5.2) and using equation (4.14) yields

$$X(s) = -M(s)\frac{G(s)}{M(s)N(s)L(s)[F(s)G(s) - E(s)H(s)]}N(s)L(s)F(s).$$

Simplifying this yields

$$X(s) = \frac{1}{\dfrac{E(s)H(s)}{F(s)G(s)} - 1}. \tag{5.7}$$

For periodic waves ($s = \mathrm{j}\omega$) the above yields

$$X(\mathrm{j}\omega) = \frac{1}{Q(\omega)\mathrm{e}^{\mathrm{j}\phi(\omega)} - 1} \tag{5.8}$$

where $Q(\omega)$ and $\phi(\omega)$ denote the magnitude and phase, respectively, of the ratio $E(\mathrm{j}\omega)H(\mathrm{j}\omega)/[F(\mathrm{j}\omega)G(\mathrm{j}\omega)]$:

$$\frac{E(\mathrm{j}\omega)H(\mathrm{j}\omega)}{F(\mathrm{j}\omega)G(\mathrm{j}\omega)} = Q(\omega)\,\mathrm{e}^{\mathrm{j}\phi(\omega)}. \tag{5.9}$$

Simplification of equation (5.8) yields the magnitude $B(\omega)$ and phase $\theta(\omega)$ of $X(\mathrm{j}\omega)$ as

$$B(\omega) = [Q^2(\omega) + 1 - 2Q(\omega)\cos\phi(\omega)]^{-1/2}$$

$$\theta(\omega) = \tan^{-1}\left[\frac{Q(\omega)\sin\phi(\omega)}{1 - Q(\omega)\cos\phi(\omega)}\right] + 2m\pi; \qquad m = 0, \pm 1, \ldots. \tag{5.10}$$

In order to relate the stability of the system to the geometrical arrangement of its components in the medium, let the primary and secondary sources be located in a three-dimensional UVW-space at points $P(0, 0, 0)$ and $S(u_s, v_s, w_s)$, respectively, with a distance d relative to each other, the detector at point $D(u_d, v_d, w_d)$ with distances r_e and r_f relative to the primary and secondary

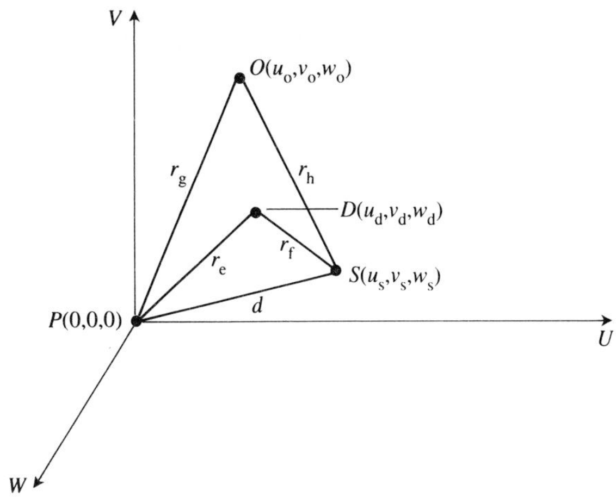

Fig. 5.2. Geometrical arrangement of the ANC system in three-dimensional coordinates

sources, respectively, and the observer at point $O(u_o, v_o, w_o)$ with distances r_g and r_h relative to the primary and secondary sources, respectively (see Fig. 5.2). The distances d, r_e, r_f, r_g, and r_h are thus given by

$$\begin{aligned} r_e &= [u_d^2 + v_d^2 + w_d^2]^{1/2} \\ r_f &= [(u_d - u_s)^2 + (v_d - v_s)^2 + (w_d - w_s)^2]^{1/2} \end{aligned} \tag{5.11}$$

$$\begin{aligned} r_g &= [u_o^2 + v_o^2 + w_o^2]^{1/2} \\ r_h &= [(u_o - u_s)^2 + (v_o - v_s)^2 + (w_o - w_s)^2]^{1/2} \end{aligned} \tag{5.12}$$

and

$$d = [u_s^2 + v_s^2 + w_s^2]^{1/2}. \tag{5.13}$$

The transfer functions $E(j\omega)$, $F(j\omega)$, $G(j\omega)$ and $H(j\omega)$ are accordingly given in terms of the distances r_e, r_f, r_g, and r_h, as in equations (4.16). Substituting for these from equation (4.16) into equation (5.9) and simplifying yields $Q(\omega)$ and $\phi(\omega)$ as

$$Q(\omega) = \frac{r_f r_g}{r_e r_h}$$

$$\phi(\omega) = \frac{2\pi}{\lambda} [(r_h - r_g) - (r_f - r_e)]$$

or

$$Q(\omega)=\frac{\left(\dfrac{r_g}{r_h}\right)}{\left(\dfrac{r_e}{r_f}\right)} \tag{5.14}$$

$$\phi(\omega)=\frac{2\pi}{\lambda}(r_{gh}-r_{ef})$$

where $r_{gh}=r_h-r_g$ and $r_{ef}=r_f-r_e$.

It follows from equations (5.8) and (5.14) that a knowledge of only the distances r_e, r_f, r_g, r_h and the signal wavelength is sufficient to describe the transfer function $X(j\omega)$ completely. Therefore, using these relations the stability of the ANC system can be determined in terms of the locations of the detector and observer relative to the primary and secondary sources in the medium.

5.4.1 Gain margin

Substituting for $\theta(\omega)$ from equations (5.10) into equation (5.5) yields

$$\tan^{-1}\left[\frac{Q(\omega)\sin\phi(\omega)}{1-Q(\omega)\cos\phi(\omega)}\right]+2m\pi=-\pi; \qquad m=0,\pm 1,\ldots$$

or

$$\frac{Q(\omega)\sin\phi(\omega)}{Q(\omega)\cos\phi(\omega)-1}=\tan(2m+1)\pi, \qquad m=0,\pm 1,\ldots. \tag{5.15}$$

It follows from equation (5.15) that the following conditions should hold

$$Q(\omega)\sin\phi(\omega)=0$$

$$Q(\omega)\cos\phi(\omega)-1<0.$$

Since $Q(\omega)$ is a strictly positive real number the above equations further lead to

$$\begin{aligned}&\sin\phi(\omega)=0, \quad \cos\phi(\omega)=+1 \quad \text{and} \quad Q(\omega)<1\\ &\sin\phi(\omega)=0, \quad \cos\phi(\omega)=-1 \quad \text{and} \quad Q(\omega)>0\end{aligned} \tag{5.16}$$

where, in the second of equations (5.16), $Q(\omega) > -1$ has been replaced by $Q(\omega) > 0$ to correspond to its realistic range.

Simplifying equations (5.16) yields the angle $\phi(\omega)$ as

$$\phi(\omega) = \begin{cases} 2n\pi & \text{for } Q(\omega) < 1 \\ (2n+1)\pi & \text{for } Q(\omega) > 0 \end{cases}; \qquad n = 0, \pm 1, \ldots$$

or

$$\phi(\omega) = \begin{cases} (2n+1)\pi & \text{for } Q(\omega) \geq 1 \\ n\pi & \text{for } 0 < Q(\omega) < 1 \end{cases}; \qquad n = 0, \pm 1, \ldots. \tag{5.17}$$

This implies that $\phi(\omega)$ is to be an integer multiple of π. Substituting for $\phi(\omega)$ from equations (5.14) into equation (5.17) and simplifying yields

$$r_{\mathrm{gh}} - r_{\mathrm{ef}} = \begin{cases} (2n+1)\dfrac{\lambda}{2} & \text{for } Q(\omega) \geq 1 \\ n\left(\dfrac{\lambda}{2}\right) & \text{for } 0 < Q(\omega) < 1 \end{cases}; \qquad n = 0, \pm 1, \ldots. \tag{5.18}$$

Equations (5.17) and (5.18) give the necessary conditions under which the phase $\theta(\omega)$ of the transfer function $X(\mathrm{j}\omega)$ is -180°. In such a situation, the gain margin k_g is given by equation (5.5).

If $\phi(\omega)$ in equation (5.17) is an even multiple of π $(Q(\omega) < 1)$ then equations (5.5) and (5.10) yield the gain margin as

$$k_g = [Q^2(\omega) + 1 - 2Q(\omega)]^{1/2} = |Q(\omega) - 1|; \qquad 0 < Q(\omega) < 1$$

or

$$k_g = 1 - Q(\omega) \qquad \text{for } 0 < Q(\omega) < 1 \qquad \text{and}$$

$$\phi(\omega) = 2n\pi;\ n = 0, \pm 1, \ldots. \tag{5.19}$$

If $\phi(\omega)$ is an odd multiple of π $(Q(\omega) > 0)$ then equations (5.5) and (5.10) yield the gain margin as

$$k_g = [Q^2(\omega) + 1 + 2Q(\omega)]^{1/2} = |Q(\omega) + 1|; \qquad Q(\omega) > 0$$

or

$$k_g = Q(\omega) + 1 \qquad \text{for } Q(\omega) > 0$$

$$\text{and} \qquad \phi(\omega) = (2n+1)\pi;\ n = 0, \pm 1, \ldots. \tag{5.20}$$

Therefore, under the conditions in equations (5.17) and (5.18), the gain margin is given by

$$k_g = \begin{cases} 1 - Q(\omega) & \text{for } \phi(\omega) = 2n\pi \text{ and } 0 < Q(\omega) < 1 \\ 1 + Q(\omega) & \text{for } \phi(\omega) = (2n+1)\pi \text{ and } Q(\omega) > 0 \end{cases} \tag{5.21}$$

where $n = 0, \pm 1, \ldots$.

Using the Nyquist stability criterion given by equation (5.4) it may be concluded that for stable operation of the system the gain margin should assume values greater than unity:

$$k_g > 1. \tag{5.22}$$

To find specifically the regions in the three-dimensional UVW-space of Fig. 5.2 corresponding to equation (5.21) and analyse system stability in these regions the two possible cases of $Q(\omega) \geq 1$ and $Q(\omega) < 1$ are considered.

$Q(\omega)$ greater than or equal to unity Substituting for $Q(\omega) \geq 1$ into equation (5.14) and simplifying yields

$$\frac{r_e}{r_f} \leq \frac{r_g}{r_h}$$

or

$$\frac{r_e}{r_f} \leq a \tag{5.23}$$

where a is the distance ratio r_g/r_h.

$$a = \frac{r_g}{r_h}. \tag{5.24}$$

It follows from Appendix A that the distance ratio in equation (5.24), for a as a parameter, defines a family of spheres in the three-dimensional UVW-space of Fig. 5.2. Substituting for r_g and r_h from equations (5.12) into equation (5.24), using equation (5.13) and simplifying yields the locus of points $O(u_o, v_o, w_o)$ in Fig. 5.2 describing this family as

$$\left[u_o + \frac{a^2 u_s}{1-a^2}\right]^2 + \left[v_o + \frac{a^2 v_s}{1-a^2}\right]^2 + \left[w_o + \frac{a^2 w_s}{1-a^2}\right]^2 = \left[\frac{ad}{1-a^2}\right]^2; \quad a \neq 1. \tag{5.25}$$

This, for a distance ratio a equal to zero, corresponds to point P (location of the primary source) and, for an infinitely large distance ratio, to point S (location of the secondary source). If the distance ratio is equal to unity then equation (5.24) defines a plane surface that perpendicularly bisects the line PS

joining the locations of the primary and secondary sources. This is described by

$$\frac{u_o}{\left(\frac{d^2}{2u_s}\right)} + \frac{v_o}{\left(\frac{d^2}{2v_s}\right)} + \frac{w_o}{\left(\frac{d^2}{2w_s}\right)} = 1. \tag{5.26}$$

It follows from the above that for $Q(\omega) \geq 1$ the detection point in Fig. 5.2 should remain inside the sphere defined by equation (5.25). This, in terms of the distance ratio a, means that for $a < 1$ the locus of points $D(u_d, v_d, w_d)$ in Fig. 5.2 should define the region on or inside the sphere in equation (5.25); for $a = 1$ point D should be on and in the region on the side of the plane in equation (5.26) that contains point P; and for $a > 1$ point D should be on or outside the sphere defined by equation (5.25). Substituting for r_e and r_f from equations (5.11) into equation (5.23), using equation (5.13), and simplifying yields

$$\left[u_d + \frac{a^2 u_s}{1 - a^2}\right]^2 + \left[v_d + \frac{a^2 v_s}{1 - a^2}\right]^2 + \left[w_d + \frac{a^2 w_s}{1 - a^2}\right]^2 \leq \left[\frac{ad}{1 - a^2}\right]^2; \quad a < 1$$

$$\frac{u_d}{\left(\frac{d^2}{2u_s}\right)} + \frac{v_d}{\left(\frac{d^2}{2v_s}\right)} + \frac{w_d}{\left(\frac{d^2}{2w_s}\right)} \leq 1; \quad a = 1 \tag{5.27}$$

$$\left[u_d - \frac{a^2 u_s}{a^2 - 1}\right]^2 + \left[v_d - \frac{a^2 v_s}{a^2 - 1}\right]^2 + \left[w_d - \frac{a^2 w_s}{a^2 - 1}\right]^2 \geq \left[\frac{ad}{a^2 - 1}\right]^2; \quad a > 1.$$

Regions for the detector (shaded) and observer corresponding to the above are marked in two dimensions in Fig. 5.3. Revolution of these diagrams around the line passing through points P and S lead to the corresponding three-dimensional descriptions.

It follows from the above discussion that if the observer O and detector D are restricted to the regions of the UVW-space defined by equations (5.25) and (5.27), respectively, then the gain $Q(\omega)$ will be greater than or equal to unity. This, in terms of equations (5.18) and (5.21), means that for such observation and detection points where the distance difference $r_{gh} - r_{ef}$ is an odd multiple of half the signal wavelength (implying that the angle $\phi(\omega)$ is an odd multiple of π and the angle $\theta(\omega)$ in equation (5.3) is $-180°$) the gain margin k_g of the system in Fig. 5.1 is given by equation (5.21) as

$$k_g = 1 + Q(\omega) \qquad \text{for } \phi(\omega) = (2n + 1)\pi \text{ and } Q(\omega) \geq 1; n = 0, \pm 1, \ldots. \tag{5.28}$$

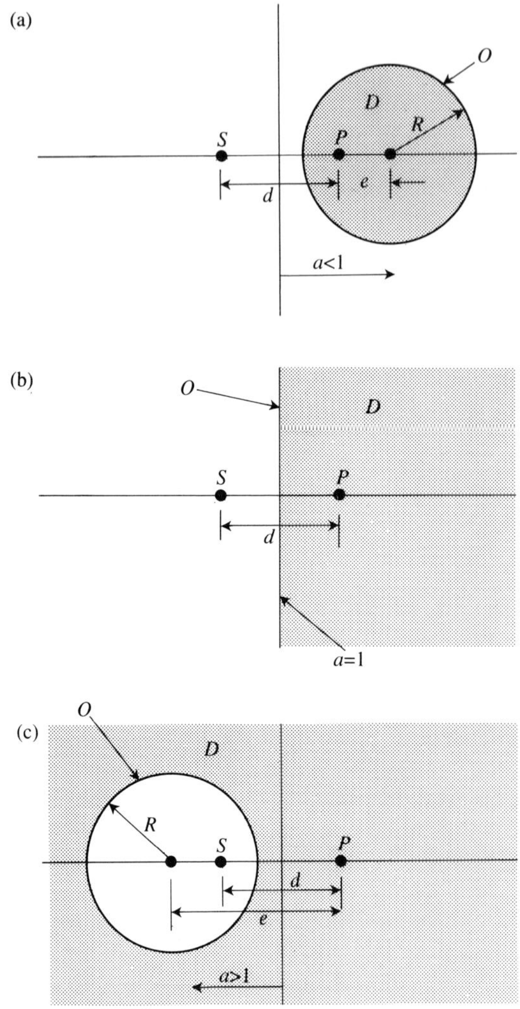

$$\left[e = \frac{a^2 d}{|a^2 - 1|} \quad \text{and} \quad R = \frac{e}{a} \right]$$

Fig. 5.3. Regions of observation point O and detection point D for $Q \geq 1$

It follows from equation (5.28) that in this situation the gain margin assumes values that are either equal to or greater than 2:

$$k_g \geq 2 \qquad \text{for } \phi(\omega) = (2n + 1)\pi \text{ and } Q(\omega) \geq 1. \tag{5.29}$$

It follows from the above and the condition for stability of the system given in equation (5.22) that, if the observer and detector are restricted to the regions defined by equations (5.25) and (5.27) of the three-dimensional UVW-space of Fig. 5.2, then at locations for which $\phi(\omega)$ is an odd multiple of π (the feedback loop in Fig. 5.1 having a phase $\theta(\omega) = -180°$) the system will be stable.

$Q(\omega)$ less than unity Substituting for $Q(\omega) < 1$ into equation (5.14) and using equation (5.24) yields

$$\frac{r_e}{r_f} > a. \tag{5.30}$$

It follows from the results obtained above for $Q(\omega) \geq 1$ that equation (5.30) for points of observation O and detection D defines a region of the three-dimensional UVW-space in Fig. 5.2 that is the corresponding complement of the regions given in equations (5.25) and (5.27), that is, for the observer restricted to the locus defined by equation (5.25) the detector is to be restricted to the following

$$\begin{aligned}
&\left[u_d + \frac{a^2 u_s}{1-a^2}\right]^2 + \left[v_d + \frac{a^2 v_s}{1-a^2}\right]^2 + \left[w_d + \frac{a^2 w_s}{1-a^2}\right]^2 > \left[\frac{ad}{1-a^2}\right]^2; \quad a < 1\\
&\frac{u_d}{\left(\dfrac{d^2}{2u_s}\right)} + \frac{v_d}{\left(\dfrac{d^2}{2v_s}\right)} + \frac{w_d}{\left(\dfrac{d^2}{2w_s}\right)} > 1; \quad a = 1 \qquad (5.31)\\
&\left[u_d - \frac{a^2 u_s}{a^2-1}\right]^2 + \left[v_d - \frac{a^2 v_s}{a^2-1}\right]^2 + \left[w_d - \frac{a^2 w_s}{a^2-1}\right]^2 < \left[\frac{ad}{a^2-1}\right]^2; \quad a > 1.
\end{aligned}$$

This is shown in two dimensions in Fig. 5.4 where the shaded areas indicate allowable locations of the detection point D. Revolution of these diagrams around the line passing through points P and S lead to the corresponding three-dimensional descriptions.

If the observer and detector are restricted to their respective regions defined by equations (5.25) and (5.31) then at locations for which the distance difference $r_{gh} - r_{ef}$ is an integral multiple of the signal wavelength (implying that the angle $\phi(\omega)$ is an integral multiple of π for which the phase $\theta(\omega)$ in equation (5.3) will be $-180°$) the gain margin of the system in Fig. 5.1 as given by equation (5.21) is

$$k_g = \begin{cases} 1 - Q(\omega) & \text{for } \phi(\omega) = 2n\pi \\ 1 + Q(\omega) & \text{for } \phi(\omega) = (2n+1)\pi \end{cases}; \qquad n = 0, \pm 1, \ldots \tag{5.32}$$

from which the ranges of the gain margin are obtained as

$$\begin{matrix} k_g < 1 & \text{for } \phi(\omega) = 2n\pi \\ k_g > 1 & \text{for } \phi(\omega) = (2n+1)\pi \end{matrix}; \qquad n = 0, \pm 1, \ldots. \tag{5.33}$$

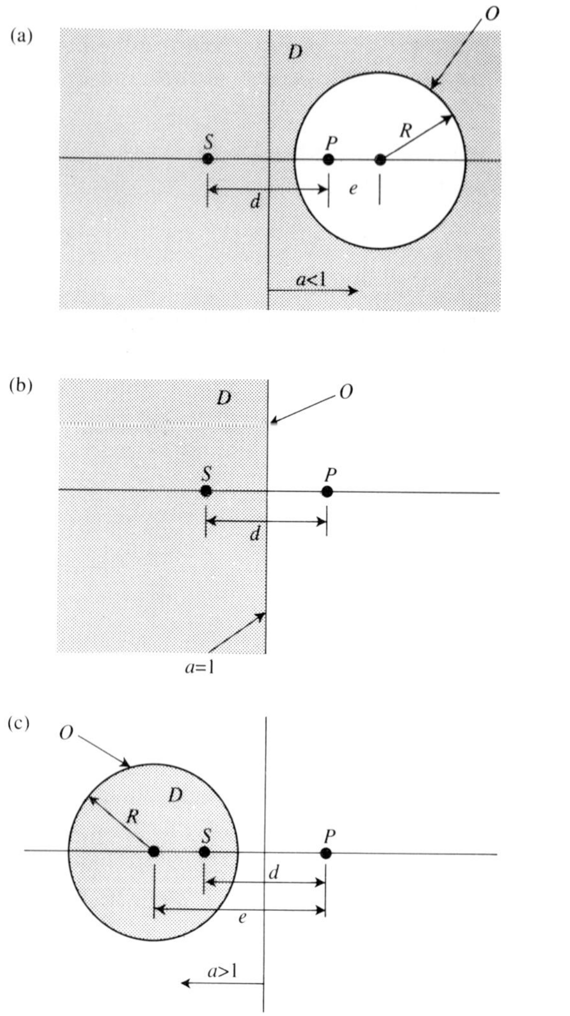

$$\left[e = \frac{a^2 d}{|a^2 - 1|} \quad \text{and} \quad R = \frac{e}{a} \right]$$

Fig. 5.4. Regions of observation point O and detection point D for $Q < 1$

It follows from equations (5.33) and (5.22) that for such detection and observation points in the region of the three-dimensional UVW-space defined by equation (5.30) where $\phi(\omega)$ is an even multiple of π the gain margin of the feedback loop in Fig. 5.1 is less than unity and hence the system will be unstable. However, if at such points $\phi(\omega)$ is an odd multiple of π then the gain margin is greater than unity and hence the system will be stable.

5.4.2 *Phase margin*

The phase margin, k_θ, of the feedback loop in Fig. 5.1 is given by equation (5.6). Substituting for $B(\omega)$ from equation (5.10) into equation (5.6) yields

$$[Q^2(\omega) + 1 - 2Q(\omega)\cos\phi(\omega)]^{-1/2} = 1. \tag{5.34}$$

Simplifying this yields

$$\cos\phi(\omega) = \frac{Q(\omega)}{2}. \tag{5.35}$$

The amplitude $Q(\omega)$ is a strictly positive quantity. Moreover, in order for $\phi(\omega)$ to have a realistic value satisfying equation (5.35) $Q(\omega)$ should not exceed 2:

$$0 < Q(\omega) \leq 2. \tag{5.36}$$

Equations (5.35) and (5.36) yield the range of $\phi(\omega)$ as

$$2n\pi - \frac{\pi}{2} < \phi(\omega) < 2n\pi + \frac{\pi}{2} \qquad \text{for } 0 < Q(\omega) \leq 2;\ n = 0, \pm 1, \ldots$$

or

$$\frac{(4n-1)\pi}{2} < \phi(\omega) < \frac{(4n+1)\pi}{2} \qquad \text{for } 0 < Q(\omega) \leq 2;\ n = 0, \pm 1, \ldots. \tag{5.37}$$

Substituting for $\phi(\omega)$ from equation (5.14) into equation (5.37) and simplifying yields the distance difference $r_{gh} - r_{ef}$ as

$$(4n-1)\frac{\lambda}{4} < r_{gh} - r_{ef} < (4n+1)\frac{\lambda}{4} \qquad \text{for } 0 < Q(\omega) \leq 2;\ n = 0, \pm 1, \ldots. \tag{5.38}$$

Equation (5.38), or equivalently equation (5.37), is the necessary condition for $B(\omega) = 1$ in equation (5.6). Under this condition the angle $\theta(\omega)$ follows from equation (5.10). Using equations (5.35) and (5.37) the following is obtained

$$\sin\phi(\omega) = \begin{cases} \frac{1}{2}\sqrt{4 - Q^2(\omega)} & \text{for } 0 \leq \phi(\omega) < \frac{(4n+1)\pi}{2} \\ -\frac{1}{2}\sqrt{4 - Q^2(\omega)} & \text{for } \frac{(4n-1)\pi}{2} < \phi(\omega) < 0 \end{cases} \tag{5.39}$$

where $0 < Q(\omega) \leq 2$ and $n = 0, \pm 1, \ldots$. Substituting for $\cos\phi(\omega)$ and $\sin\phi(\omega)$ from equations (5.35) and (5.39) into equation (5.10) and simplifying yields the phase angle $\theta(\omega)$ as

$$\theta(\omega) = \begin{cases} \tan^{-1}\dfrac{Q(\omega)\sqrt{4 - Q^2(\omega)}}{2 - Q^2(\omega)} + 2m\pi & \text{for } 0 \leq \phi(\omega) < \dfrac{(4n+1)\pi}{2} \\ -\tan^{-1}\dfrac{Q(\omega)\sqrt{4 - Q^2(\omega)}}{2 - Q^2(\omega)} + 2m\pi & \text{for } \dfrac{(4n-1)\pi}{2} < \phi(\omega) < 0 \end{cases} \tag{5.40}$$

where $0 < Q(\omega) \leq 2$, $m = 0, \pm 1, \ldots$ and $n = 0, \pm 1, \ldots$. This is shown as a function of $Q(\omega)$ for $m = 0$ in Fig. 5.5(a).

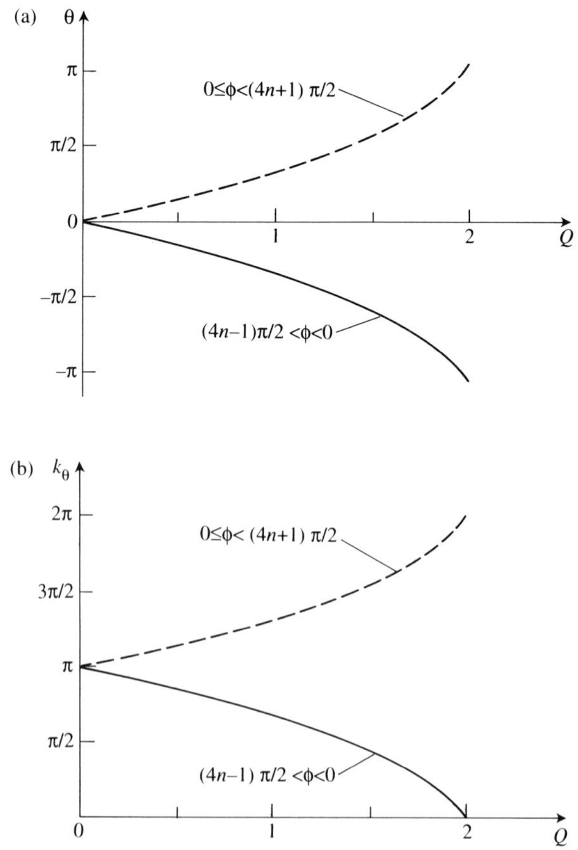

Fig. 5.5. (a) Phase, (b) phase margin of the feedback loop

Substituting for $\theta(\omega)$ from equation (5.40) into equation (5.6) yields the gain margin as

$$k_\theta = \begin{cases} \tan^{-1}\dfrac{Q(\omega)\sqrt{4-Q^2(\omega)}}{2-Q^2(\omega)} + (2m+1)\pi & \text{for } 0 \le \phi(\omega) < \dfrac{(4n+1)\pi}{2} \\ -\tan^{-1}\dfrac{Q(\omega)\sqrt{4-Q^2(\omega)}}{2-Q^2(\omega)} + (2m+1)\pi & \text{for } \dfrac{(4n-1)\pi}{2} < \phi(\omega) < 0 \end{cases} \tag{5.41}$$

where $0 < Q(\omega) \le 2$ and m and n are integer numbers. This is shown as function of $Q(\omega)$ for $m = 0$ in Fig. 5.5(b). To find the corresponding region of the three-dimensional UVW-space of Fig. 5.2 in terms of the locations of the observation and detection points we consider equation (5.36).

Substituting for $Q(\omega)$ from equation (5.14) into equation (5.36), using equation (5.24), and simplifying yields

$$\frac{r_e}{r_f} \ge 0.5\,a. \tag{5.42}$$

Substituting for r_e and r_f from equations (5.11) into equation (5.42), simplifying, and using equation (5.13) yields the locus of points $D(u_d, v_d, w_d)$ as

$$\left[u_d + \frac{a^2 u_s}{2-a^2}\right]^2 + \left[v_d + \frac{a^2 v_s}{2-a^2}\right]^2 + \left[w_d + \frac{a^2 w_s}{2-a^2}\right]^2 \ge \left[\frac{\sqrt{2}ad}{2-a^2}\right]^2; \qquad a < 2$$

$$\frac{u_d}{\left(\dfrac{d^2}{2u_s}\right)} + \frac{v_d}{\left(\dfrac{d^2}{2v_s}\right)} + \frac{w_d}{\left(\dfrac{d^2}{2w_s}\right)} \ge 1; \qquad a = 2 \tag{5.43}$$

$$\left[u_d - \frac{a^2 u_s}{a^2-2}\right]^2 + \left[v_d - \frac{a^2 v_s}{a^2-2}\right]^2 + \left[w_d - \frac{a^2 w_s}{a^2-2}\right]^2 < \left[\frac{\sqrt{2}ad}{a^2-2}\right]^2; \qquad a > 2.$$

These are shown in two dimensions in Fig. 5.6 revolution of which around the line passing through points P and S leads to the corresponding three-dimensional description.

It follows from the above analysis that, for the detector and observer locations in the region of the three-dimensional UVW-space of Fig. 5.2 defined by equation (5.36), or equivalently by equations (5.43), for which equation (5.35) holds, the amplitude $B(\omega)$ in equation (5.3) is unity and in such a case the phase margin k_θ of the feedback loop in Fig. 5.1 is given by equation (5.41). In this case, for a minimum-phase condition the system will have stable operation at locations of detector and observer in the regions defined by equations (5.43) where the phase margin assumes positive values, whereas for negative values of k_θ the system will be unstable.

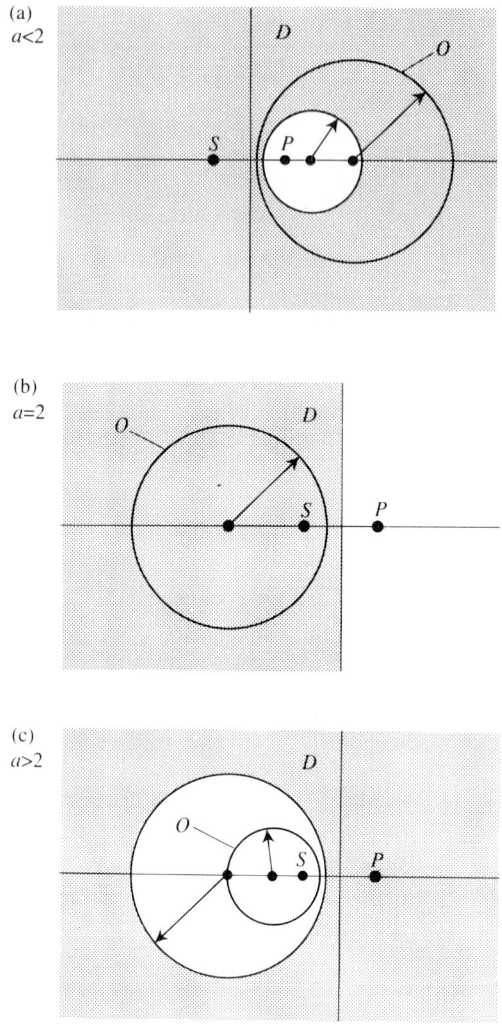

Fig. 5.6. Regions of observation point O and detection point D for $Q \leq 2$

5.5 Summary

A stability analysis of the ANC system has been presented. The detector in an ANC system gives a combined measure of the primary and secondary signals reaching the detection point. The acoustic path between the secondary source and detector thus forms a closed feedback loop with the controller path. Therefore, the secondary source radiation entering the detector via this acoustic feedback path can cause the system to become unstable.

The use of unidirectional detectors and sources, multiple-detector multiple-source configurations, or indirect detection will lead to isolating (to some degree) the detector from secondary source radiation and thus avoid the problem of instability in specific situations. However, a stability analysis of the system results in a robust ANC system design. For practical systems the use of absolute stability is not beneficial. In this respect a measure of the relative stability in terms of gain and phase margins provides a more acceptable design criterion.

The dependence of controller characteristics on the characteristics of system components and geometry leads to the dependence of the stability and relative stability of the system on geometric composition of the detection and observation points. Certain combinations of the observation and detection points in the medium can cause the system to become unstable. These points describe regions of the medium as joint loci of detection and observation points leading to instability in system response. Such regions are identified through a mapping of stability conditions on to the system geometry, thus leading to practical constraints on geometric composition of the observation and detection points.

5.6 References

D'Azzo, J.J. and Houpis, C.H. (1966). *Feedback control analysis and synthesis*, (2nd ed). McGraw-Hill, New York.

Dines, P.J. (1982). Comparison of least squares estimation and impulse response techniques for active control of flame noise. *Proceedings of the Institute of Acoustics*, November, F2.1–F2.4.

Jessel, M. and Mangiante, G.A. (1972). Active sound absorbers in an air duct. *Journal of Sound and Vibration*, **23**, (3), 383–90.

Kim, C.D., Morikawa, S., Abe, M., and Kido, K. (1984). Active suppression of noise in microphone line using vibration of noise source. *Proceedings of Inter-noise 84: International Conference on Noise Control Engineering*, Honolulu, **I**, 469–72.

Leventhal, H.G. and Eghtesadi, Kh. (1979). Active attenuation of noise: monopole and dipole systems. *Proceedings of Inter-noise 79: International Conference on Noise Control Engineering*, Warsaw, **I**, 175–80.

Ross, C.F. (1982). Quietening newspaper printing presses by active control. *Proceedings of the Institute of Acoustics*, November, F1.1–F1.4.

Ross, C.F. (1986). Novel applications of active control techniques. *Proceedings of the Institute of Acoustics*, **8**, Part 1, 127–33.

Roure, A. (1985). Self-adaptive broadband active sound control system. *Journal of Sound and Vibration*, **101**, (3), 429–41.

Swinbanks, M.A. (1973). Active control of sound propagation in long ducts. *Journal of Sound and Vibration*, **27**, (3), 411–36.

6

DESIGN OF AN ANC SYSTEM WITH A FIXED CONTROLLER

6.1 Introduction

The resulting controller transfer function obtained from the design procedure of Chapter 4 can be implemented using either analogue or digital techniques. This consists of the process of measuring the required controller transfer function, realizing this as a continuous frequency-dependent transfer function and implementing as an analogue or digital controller. The processes of measurement and realization involve estimating the various transfer functions that appear in the controller design equation using a suitable measurement technique, calculating the controller transfer function using these measurements, and finding an appropriate transfer function that matches the controller transfer function to within practical limits. This will result in a transfer function in the continuous-time domain. The controller transfer function thus obtained can then be realized either as an analogue controller, or converted into the discrete-time domain and implemented as a digital controller.

Due to the available electronic technology, ANC systems were previously only implemented using analogue techniques which led to the various problems associated with these attempts as discussed in Chapter 2. However, the development of microprocessor technology now provides the opportunity to develop practical ANC systems. In particular, powerful digital signal processors especially designed for signal processing and digital filtering applications are now available. These have the capability for real-time processing of signals within the frequency range suitable for ANC, and are cheap enough to allow dedication to one system, thus utilizing the various advantages a digital controller has over an analogue controller.

In this chapter a procedure for the design and real-time implementation of a digital controller for ANC systems using the FFCS is given. This is based on specified observation points with suitable electronic transducers. The performance of the ANC system thus obtained is investigated under laboratory and practical environments. The exhaust noise of a motorcycle is used as a typical example of a practical source of noise.

6.2 The controller transfer function

In this section a procedure for realizing the controller as a digital controller for a particular system geometry is presented. Assume a compact source is given that emits noise of broadband nature into a three-dimensional non-dispersive propagation medium. It is desired that, with a compact cancelling source placed close to the unwanted source, the noise is cancelled at and in the vicinity of points equidistant from the two sources. This, with a loudspeaker as the cancelling source, will include part of a plane surface (in front of the loudspeaker) that perpendicularly bisects a line joining the location of the sources. If the separation between the sources is kept to a minimum the cancellation zone will cover a listener's position facing both the primary and secondary sources.

The above situation in terms of the ANC system in Figure. 4.3 leads to the equality of the distances r_g and r_h. This, in terms of the transfer functions $G(s)$ and $H(s)$ for $s = j\omega$, leads to

$$G(\mathrm{j}\omega) = H(\mathrm{j}\omega); \qquad r_g = r_h. \tag{6.1}$$

Substituting for $H(\mathrm{j}\omega)$ from equation (6.1) into equation (4.14) with $s = \mathrm{j}\omega$ yields

$$\Delta(\mathrm{j}\omega) = F(\mathrm{j}\omega)G(\mathrm{j}\omega) - E(\mathrm{j}\omega)G(\mathrm{j}\omega)$$

or

$$\Delta(\mathrm{j}\omega) = G(\mathrm{j}\omega)[F(\mathrm{j}\omega) - E(\mathrm{j}\omega)]. \tag{6.2}$$

Substituting for $\Delta(\mathrm{j}\omega)$ from equation (6.2) into equation (4.13) and simplifying yields the controller transfer function as

$$C(s) = \frac{1}{M(s)N(s)L(s)[F(s) - E(s)]}; \qquad G(s) = H(s), \qquad s = \mathrm{j}\omega. \tag{6.3}$$

This is the required controller transfer function for cancellation of noise at and in the vicinity of observation points that are equidistant from the two sources. This can thus be expressed as

$$C(\mathrm{j}\omega) = \frac{1}{X_2(\mathrm{j}\omega) - X_1(\mathrm{j}\omega)} \tag{6.4}$$

where

$$\begin{aligned} X_1(\mathrm{j}\omega) &= M(\mathrm{j}\omega)N(\mathrm{j}\omega)L(\mathrm{j}\omega)E(\mathrm{j}\omega) = A_1(\omega)\,\mathrm{e}^{\mathrm{j}\theta_1(\omega)} \\ X_2(\mathrm{j}\omega) &= M(\mathrm{j}\omega)N(\mathrm{j}\omega)L(\mathrm{j}\omega)F(\mathrm{j}\omega) = A_2(\omega)\,\mathrm{e}^{\mathrm{j}\theta_2(\omega)}. \end{aligned} \tag{6.5}$$

It follows from equation (6.4) that a method of obtaining the required frequency response of the controller would be to first measure $X_1(\mathrm{j}\omega)$ and $X_2(\mathrm{j}\omega)$ for a broad frequency range of the primary wave over which cancellation is desired and then use equation (6.4) to calculate $C(j\omega)$. The desired frequency response thus obtained can then be mapped on to a transfer function that closely fits the amplitude and phase characteristics. Care must be taken in this process to ensure that the resulting amplitude and phase characteristics in relation to the required characteristics in equation (6.4) are such as to satisfy the conditions for cancellation of the primary noise over the desired frequency range. This will require the power ratio α of the secondary wave to the primary wave not to exceed $4\beta^2$:

$$0 < \alpha < 4\beta^2. \tag{6.6}$$

The power ratio α in terms of the primary signal $P_o(j\omega)$ and secondary signal $S_o(j\omega)$ is given by

$$\alpha = \frac{|S_o(j\omega)|^2}{|P_o(j\omega)|^2}.$$

This, using Fig. 4.5 and equation (6.1) yields

$$\alpha = |A(j\omega)|^2. \tag{6.7}$$

Substituting for $A(j\omega)$ from equation (4.8), with $s = j\omega$, into equation (6.7) and using equations (6.5) yields

$$\alpha = \left|\frac{X_1(j\omega)C_a(j\omega)}{1 - X_2(j\omega)C_a(j\omega)}\right|^2 \tag{6.8}$$

where $C_a(j\omega)$ represents the actual (realized) frequency response of the controller. If the allowable offset (error) in the controller characteristics relative to the required values are a factor of $a_c(\omega)$ in the amplitude and an amount of $\phi_c(\omega)$ in the phase then this is given by

$$C_a(j\omega) = C(j\omega)a_c(\omega)\, e^{j\phi_c(\omega)}. \tag{6.9}$$

Substituting for α from equation (6.8) into equation (6.6), using equations (6.5) and (6.9) and simplifying yields

$$0 < \frac{a_c^2(\omega)}{\beta^2} < \frac{A_1^2(\omega) + A_2^2(\omega)[1 - 2a_c(\omega)\cos\phi_c(\omega)]}{0.25\, A_1^2(\omega) - A_2^2(\omega)\beta^2}$$
$$- \frac{2A_1(\omega)A_2(\omega)\{\cos[\theta_1(\omega) - \theta_2(\omega)] - a_c(\omega)\cos[\theta_1(\omega) - \theta_2(\omega) - \phi_c(\omega)]\}}{0.25\, A_1^2(\omega) - A_2^2(\omega)\beta^2} \tag{6.10}$$

$$0 < a_c(\omega) < \frac{B_2(\omega) + \sqrt{B_2^2(\omega) + B_1(\omega)B_3(\omega)}}{B_1(\omega)} \tag{6.11}$$

where

$$B_1(\omega) = 0.25\ A_1^2(\omega) - A_2^2(\omega)\beta^2.$$

$$B_2(\omega) = \{A_1(\omega)A_2(\omega)\cos[\theta_1(\omega) - \theta_2(\omega) - \phi_c(\omega)] - A_2(\omega)\cos\ \phi_c(\omega)\}\beta^2.$$

$$B_3(\omega) = \{A_1^2(\omega) + A_2^2(\omega) - 2A_1(\omega)A_2(\omega)\cos[\theta_1(\omega) - \theta_2(\omega)]\}\beta^2.$$

Equation (6.11) gives the required relation between the allowable amplitude error and phase error in the controller characteristics at frequencies ω of the component waves.

6.2.1 Frequency—response measurement

The controller design calculation follows measurement of $X_1(j\omega)$ and $X_2(j\omega)$ over a broad frequency range (say, 0–500 Hz). Such a measurement can be carried out by using the set-up shown in Fig. 6.1. Two loudspeakers, one as primary source and one as secondary source, are placed close to one another (with a minimum separation between them) and a detector (microphone) is placed opposite to the primary source loudspeaker. The source amplifier is used to drive the loudspeakers and the detector amplifier is used to amplify the microphone signals to a desired level. The processor accounts for transfer characteristics of the analogue to digital (A/D) and digital to analogue (D/A) circuitry, the anti-aliasing filter and so on, of a dedicated signal processor on which the controller is to be implemented. The amplifiers and the processor in Fig. 6.1 are some of the necessary electronic components used in the controller path of the system. The equivalent transfer function of these is indicated by $N(s)$ in Fig. 4.4, or equivalently by $N(j\omega)$ in equation (6.5). $L(j\omega)$ and $M(j\omega)$ in equations (6.5) respectively represent the frequency response of the secondary source loudspeaker and the detector microphone in Fig. 6.1.

If a suitable spectrum analyser is available then the sources can be driven by a broadband noise, a pseudo-random binary sequence (PRBS), for instance, and the frequency responses $X_1(j\omega)$ and $X_2(j\omega)$ measured accordingly. Alternatively, these frequency responses can be measured at single

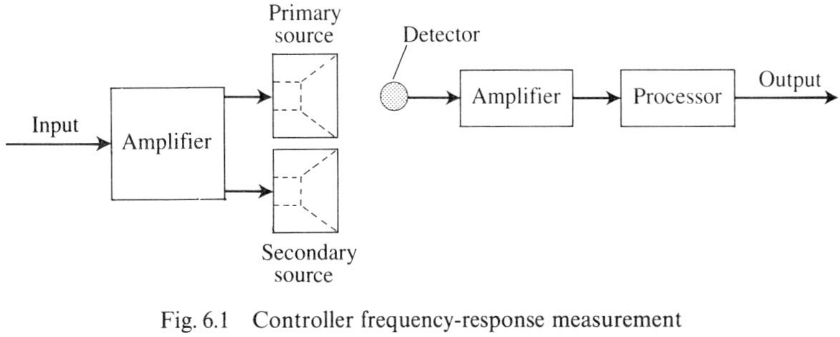

Fig. 6.1 Controller frequency-response measurement

tones over a sufficient range of frequencies of interest. Note in equation (6.5) that $M(j\omega)$, $L(j\omega)$, and $N(j\omega)$ are common to both $X_1(j\omega)$ and $X_2(j\omega)$. Therefore, depending on the transfer characteristics of the loudspeakers, two alternative procedures are suggested.

1. If the primary and secondary source loudspeakers have identical amplitude and phase characteristics over the frequency range of interest then with the set-up in Fig. 6.1:
   (a) drive the primary source loudspeaker and measure the frequency response between the input and output; this gives $X_1(j\omega)$;
   (b) drive the secondary source loudspeaker and measure the frequency response between the input and output; this gives $X_2(j\omega)$;
   (c) use the results of (a) and (b) in equation (6.4) to calculate $C(j\omega)$.
2. If the transfer characteristics of the two loudspeakers in Fig. 6.1 are not identical then:
   (a) place the secondary source loudspeaker at the location of the primary source loudspeaker, drive it and measure the frequency response between the input and output; this gives $X_1(j\omega)$;
   (b) place the secondary source loudspeaker in its original location as in Fig. 6.1, drive it and measure the frequency response between the input and output; this gives $X_2(j\omega)$;
   (c) use the results of (a) and (b) in equation (6.4) to calculate $C(j\omega)$.

Procedure 1. above relies on the assumption that the two loudspeakers have identical characteristics. In practice, this is possible to some approximate degree of accuracy. Therefore, this procedure leads to a controller transfer function that approximately matches the required transfer function. Procedure 2. leads to a more accurate and reliable controller transfer function. However, care must be taken to ensure that errors in positioning the secondary source loudspeaker at the two locations are minimized.

The location of the (omnidirectional) detector microphone relative to the primary and secondary sources in Fig. 6.1 as well as its orientation relative to the face of the primary source directly affect the resulting controller transfer function and subsequently its implementation. The distance between the detector and the primary source gives a measure of the extent of the locus of the IGC requirement. It was shown in Chapter 4 that for the region of the medium occupied by the locus of the IGC requirement to be minimized the detector must be as close to the primary source as practically possible.

The orientation of the detector microphone relative to the primary source face affects the phase characteristic of the resulting controller. Investigations conducted by the authors with the arrangement of Fig. 6.1 have shown that if the detector is placed opposite to and facing the centre of the primary source loudspeaker cone then the resulting controller will have a phase characteristic that has a minimum value at any frequency. The phase values will increase for increasing orientations from this position and will reach their maximum values when the face of the microphone is at right angles to the face of the primary source loudspeaker.

A typical set of controller transfer characteristics obtained applying procedure 2. above with the detector microphone placed opposite to and facing the primary source loudspeaker are shown by the solid-line graphs in Fig. 6.2. The controller gain appears to remain approximately flat after 100 Hz. However, below 100 Hz it increases substantially with decreasing frequencies. This is due to the correspondingly substantial attenuation in the loudspeaker with decreasing frequencies. The phase characteristic, on the other hand, starts with a value well below 0° and steadily increases for increasing frequency; the phase sweeps through an angle of 180° from 50 Hz to 500 Hz.

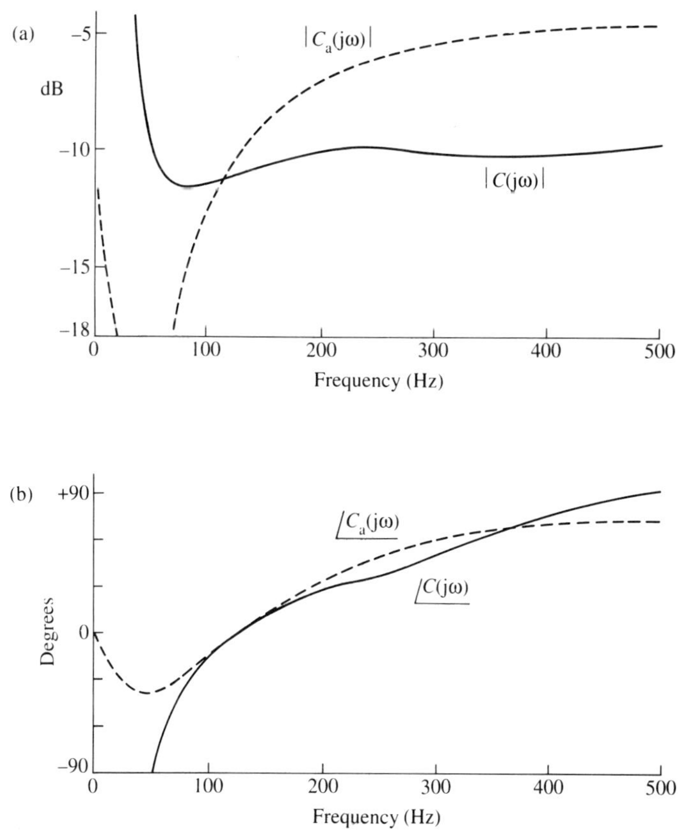

Fig. 6.2. Controller characteristics: (a) amplitude; (b) phase

6.2.2 Realization in the continuous-time domain

The controller frequency-response obtained above can now be mapped on to a suitable transfer function $C_a(s)$ in the complex frequency s domain that approximates the required amplitude and phase characteristics for $s = j\omega$. In this process care must be taken to ensure that the transfer function is stable and causal, that is, the number of its zeros must be less than or equal to the number of its poles, so that the system's stability and causality are preserved. Moreover, the practical limitations of the signal processing device on which this is to be implemented must be observed.

Some specific points to note in the process of obtaining the transfer function $C_a(s)$ are that the phase of $C_a(s)$ should be close enough to the phase of $C(s)$, to within practically acceptable limits. The amplitude characteristic of $C_a(s)$ should be of a flat shape. The level of the amplitude of $C_a(s)$ is not of much concern at this stage because it can later be boosted up by an amplifier or brought down by a potentiometer as required. Looking at the required phase characteristics in Fig. 6.2 we may choose at least one pole of $C_a(s)$ to be near the origin ($s = 0$), to account for the low phase and high gain at frequencies below 100 Hz, followed by a number of zeros nearby, so that the phase rises with frequency, and, finally, to make sure $C_a(s)$ is causal, a number of poles at frequency ranges well above 500 Hz. Note that this is based on purely real poles and zeros. Similar procedures can be adopted in terms of complex poles and zeros. These thoughts can easily be put together into a computer program to give optimum locations of poles and zeros on the basis of minimizing the error in phase characteristics and maintaining a nearly flat amplitude characteristic. A typical transfer function obtained through a computer program based on a maximum phase error of 20° ($|\phi_c(\omega)| = 20°$) over the frequency range of 80–500 Hz is given by Leitch and Tokhi (1987):

$$C_a(s) = k_s \frac{(s + 298\,\pi)(s + 300\,\pi)(s + 358\,\pi)(s + 400\,\pi)(s + 462\,\pi)}{(s + 60\,\pi)(s + 100\,\pi)(s + 1300\,\pi)(s + 1400\,\pi)(s + 1500\,\pi)} \tag{6.12}$$

where k_s is a positive real number. A maximum limit of 20° for $|\phi_c(\omega)|$ is a reasonable level of allowable error with which the condition of cancellation in equation (6.11) can be satisfied over a broad frequency range by a proper adjustment of the amplitude characteristics. The amplitude and phase of $C_a(s)$ are shown for $s = j\omega$ by the dashed-line graphs in Fig. 6.2. The corresponding amplitude error $a_c(\omega)$ and phase error $\phi_c(\omega)$ are shown in Fig. 6.3. Thus the maximum value of $|\phi_c(\omega)|$ is no greater than 20° for frequencies beyond 80 Hz. The amplitude error $a_c(\omega)$, on the other hand, can be altered, as mentioned above, by using an amplifier and/or a potentiometer in the path of the controller without affecting the phase characteristic, such that equation (6.11) holds over a broad frequency range and satisfactory cancellation is obtained. Note that with the procedure outlined above a fifth-order transfer

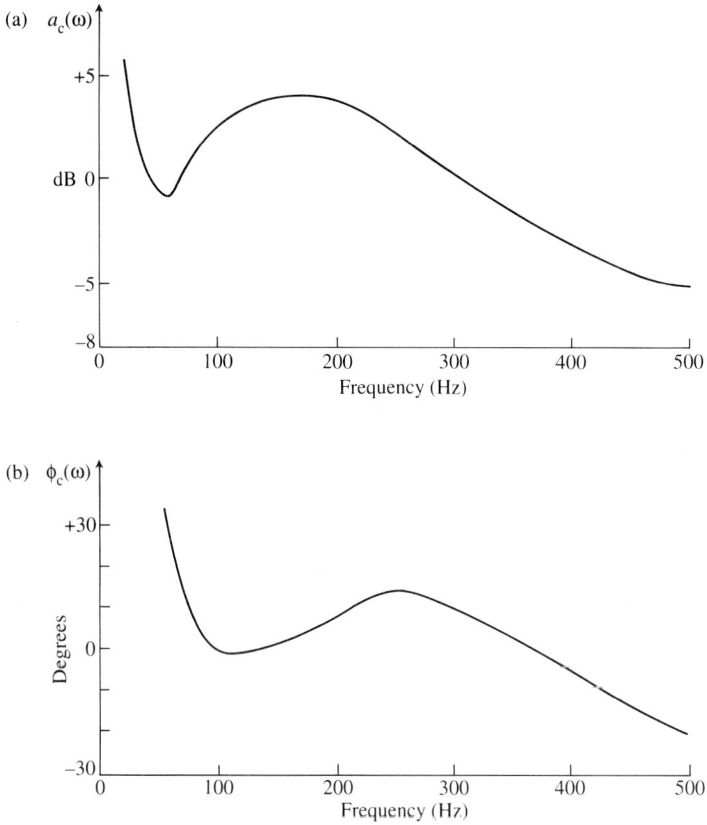

Fig. 6.3. (a) Amplitude, (b) phase error in controller characteristics

function was obtained to give a close match to the required controller transfer function. However, as will be seen later, with a more sophisticated optimization technique a transfer function of a lower order can be reached to give a similar or even closer match to the required transfer function.

6.2.3 Realization in the discrete-time domain

For purposes of discrete-time implementation, the continuous-time transfer function $C_a(s)$ in equation (6.12) is transformed to its equivalent discrete-time transfer function $C_a(z)$ through an appropriate s-plane to z-plane mapping. The discrete-time transfer function thus obtained can then be realized as a digital filter either in a recursive or non-recursive form (Antoniou 1979; Lam 1979; Oppenheim *et al.* 1984; Temes and Mitra 1973; Terrel 1980). A recursive form of realization is preferred here as the advantages of this in real-time applications are superior to those of the non-recursive realization. It is known that for the same filter specification the required order in a non-recursive design can be as high as 5–10 times that in a recursive design, depending upon the type of application. Therefore, a recursive digital filter, as compared to its non-recursive counterpart, generally uses less memory and arithmetic processing for its implementation. This further minimizes problems that may arise due to computational errors in the processor. The main disadvantage of a recursive realization is that it is possible for the filter to become unstable specifically due to poles with magnitudes close to unity. However, this problem can be avoided by keeping the poles of the transfer function inside the stability region and reasonably far away from the instability boundary.

There are a number of techniques used for transforming an analogue filter to a recursive digital filter (Antoniou 1979; Oppenheim *et al.* 1984; Temes and Mitra 1973; Terrel 1980). Among these is the bilinear z-transformation which is derived by using the trapezoidal rule for numerical integration to obtain a discrete-time equivalent to continuous-time integration (Antoniou 1979; Oppenheim *et al.* 1984; Terrel 1980). This is given by

$$s = \frac{2}{T}\frac{1 - z^{-1}}{1 + z^{-1}} \tag{6.13}$$

where T is the sampling period. Through this transformation the entire $j\omega$ axis of the s-plane is mapped into the 2π length of the unit circle in the z-plane. This means that the problem of aliasing is eliminated since the baseband is confined to half the sampling frequency, but the disadvantage of this is a distorted frequency scale, known as warping. However, if the digital filter is operated at a frequency range of less than $0.3f_s$, where f_s is the sampling frequency, then little distortion in the frequency scale will occur. Moreover, the warping effect on the amplitude response of the digital filter can be eliminated by pre-warping the analogue filter and then transforming the result to the digital filter. In the case of the phase response of the digital filter, little can be done to avoid the warping effect. The sampling period is chosen so as to minimize the distortion in the frequency scale as well as to ensure that the poles and (stable) zeros of the resulting transfer function are inside the unit circle in the z-plane. A too small value of T(that is, a too large f_s) leads to the poles tending towards $|z| = 1$, hence giving rise to unstable poles and zeros in the process of implementation due to computational errors. Moreover, if a sampling control clock is not available in the signal processor, in which case the sampling period is governed by the length of the program, then the sampling period is determined by the processor instruction cycle.

Using the bilinear z-transformation and a sampling frequency of 6.5 kHz the continuous-time transfer function $C_a(s)$ transforms to the following equivalent discrete-time transfer function

$$C_a(z) = k_z \frac{(1 - 0.8656z^{-1})(1 - 0.8648z^{-1})(1 - 0.8408z^{-1})}{(1 - 0.9714z^{-1})(1 - 0.9528z^{-1})(1 - 0.5219z^{-1})} \times \frac{(1 - 0.8237z^{-1})(1 - 0.7991z^{-1})}{(1 - 0.4944z^{-1})(1 - 0.4679z^{-1})} \quad (6.14)$$

where k_z is a positive real number. This is the required discrete-time controller transfer function and is in a form that can be implemented as a recursive digital filter using a digital signal processor.

6.2.4 *Digital implementation*

The transfer function $C_a(z)$ is realized as a digital filter using a suitable realization structure. The realization structure is chosen on the basis of avoiding the problems of overflow and minimizing the effects of quantization (as discussed later). Various realization structures of practical importance are known in digital filtering (Antoniou 1979; Oppenheim *et al.* 1984; Terrel 1980). Among these, a reliable single-input single-output (SISO) structure is the cascade structure where the transfer function is implemented in factorized form as a series of connected low-order blocks, usually of first or second order. This structure offers possibilities of optimal distribution of poles and zeros among the blocks, thereby allowing the designer to choose the internal block structures optimally.

Using a cascade filter realization structure the transfer function in equation (6.14) can be implemented as three cascade blocks of one first-order block and two second-order blocks. The constant k_z can be accounted for either within the filter structure or, if it leads to internal overflow problems, outside the filter structure by an amplifier or attenuator as necessary. Typical amplitude and phase characteristics of the filter, hereafter referred to as the digital filter controller (DFC), implemented on a digital signal processor in real time are shown in Fig. 6.4 over a linear frequency range of 0–1000 Hz. The theoretical amplitude and phase characteristics of the filter transfer function $C_a(z)$ given by equation (6.14) are also shown in Fig. 6.4 for reasons of comparison. The amplitude characteristics of both are seen to agree well, apart from a constant offset which could be compensated for by an amplifier. The phase characteristics, however, have a linear offset relative to one another. This is due to the linear phase shift introduced by the A/D and D/A circuitry of the digital signal processor. Since the phase error in the range of 0–500 Hz is still within practical limits no attempt is necessary to compensate

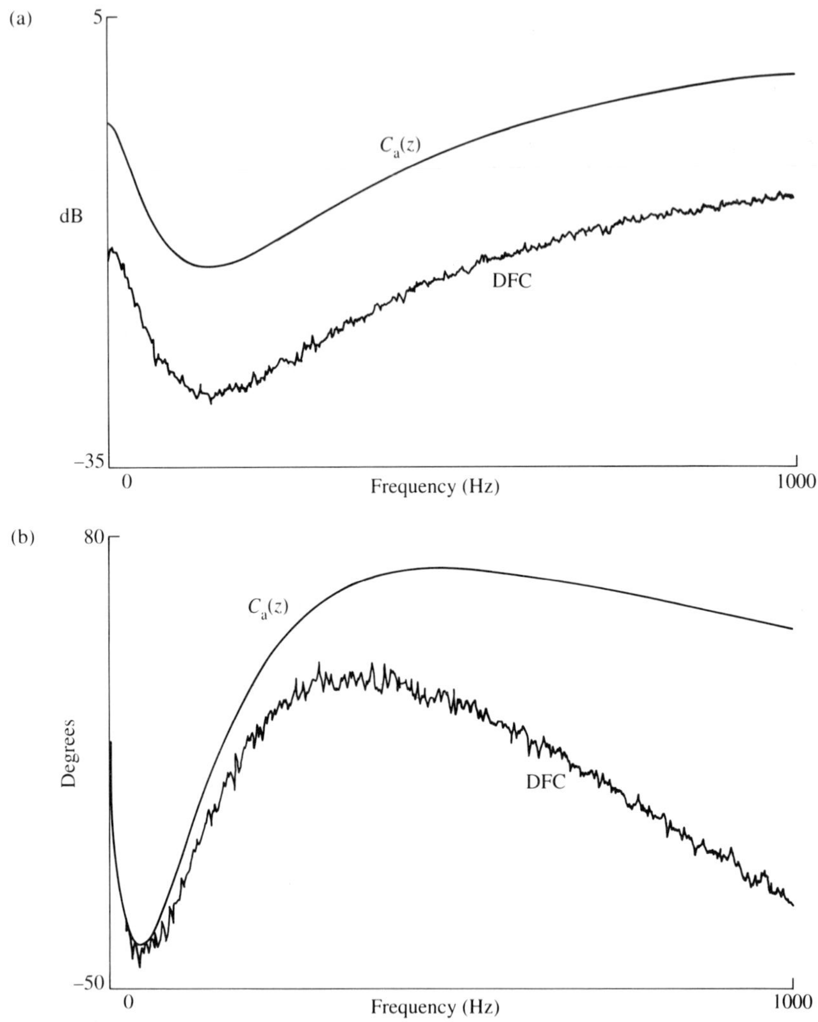

Fig. 6.4. Transfercharacteristics of DFC and $C_a(z)$: (a) amplitudes; (b) phases

for the linear phase shift. Methods of compensation for such a phase shift are discussed later in the book.

6.3 Investigation of system performance

Presented in this section are a set of experiments under laboratory and practical environments to investigate the performance of the ANC system employing the DFC as controller. More detailed discussions and further

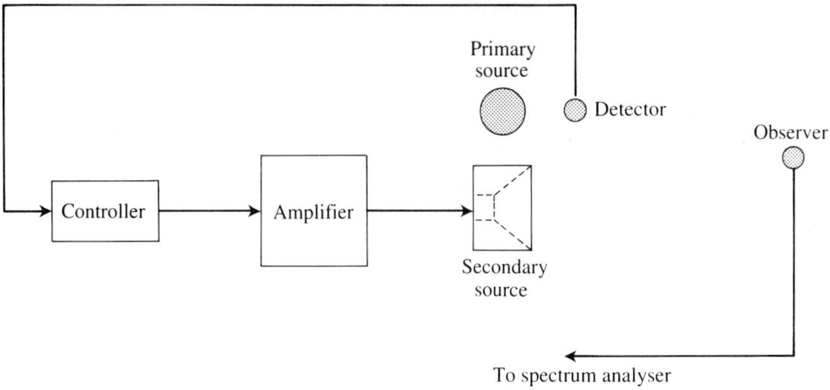

Fig. 6.5. Arrangement for active noise cancellation

results of performance investigation of the system have previously been reported by the authors (Leitch and Tokhi 1987). The experiments follow the set-up shown in Fig. 6.5 where the controller represents the DFC and the amplifier represents an audio power amplifier driving the secondary source loudspeaker. The detector microphone is situated relative to the primary and secondary sources so as to correspond to Fig. 6.1 and the observer is situated so as to be equidistant from the primary and secondary sources. A spectrum analyser is used to give a measure of the combined secondary and primary sound level at the observation point relative to the primary sound level $(G_{cco}(\omega)/G_{ppo}(\omega))$ as a function of frequency. This gives a quantitative measure of the cancellation of sound for every component frequency of the noise.

6.3.1 Synthetic source of noise

Under laboratory conditions a typical broadband white-noise source can be simulated by driving a loudspeaker by a broadband (0–500 Hz) PRBS signal. The source of noise thus obtained can be used as the unwanted primary source of noise in Fig. 6.5. In this manner a typical result of system performance, in terms of level of noise relative to the primary noise prior to cancellation in dB, as a function of frequency is shown in Fig. 6.6. This shows noise cancellation quantitatively as a function of frequency in which negative values correspond to levels of cancellation and positive values correspond to levels of reinforcement. Cancellation is seen to start below 50 Hz and extends to the maximum noise frequency of the source (500 Hz). The average level of cancellation thus achieved is more than 8 dB with the maximum peak cancellation of 23.5 dB occurring at 484 Hz. As would be expected, the cancellation in Fig. 6.6 is nearly flat. In contrast to an amplifier as a

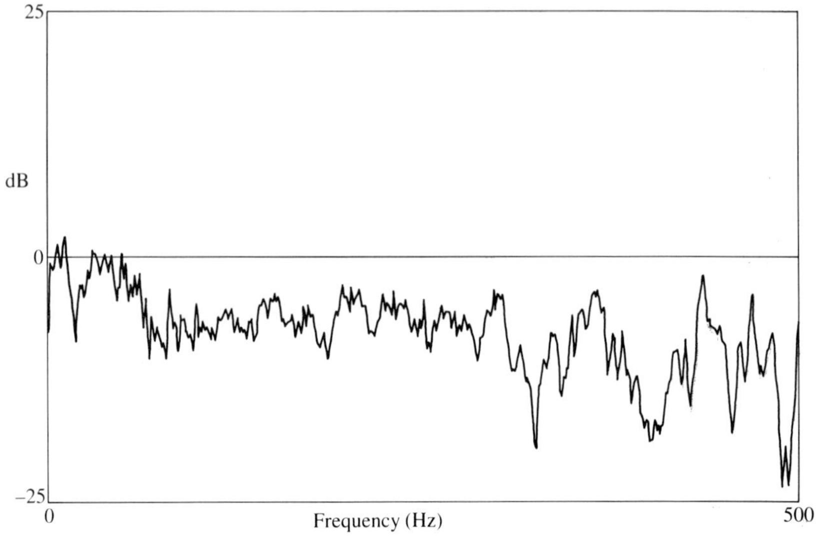

Fig. 6.6. Noise cancellation of the synthetic source

controller, which gives a nearly cosinusoidal cancellation as a function of frequency, this is due to the required frequency-dependent characteristics of the DFC. Here, the amplitude as well as phase of every tone of the detected signal is adjusted accordingly so that cancellation at each available frequency of the noise is achieved.

It follows from the amplitude and phase characteristics of the controller in Fig. 6.2 that by changing the gain level in the controller path the characteristic $|C_a(j\omega)|$ can closely approximate $|C(j\omega)|$ so that the amplitude and phase errors in the controller characteristics are within the constraints of equation (6.11) over a broad frequency range. However, due to the relatively larger offset in the amplitudes and phases of $C_a(j\omega)$ and $C(j\omega)$ at frequencies lower than 50 Hz no cancellation is obtained in that region. The relatively smaller amount of cancellation at frequencies below 300 Hz is a consequence of the larger amounts of amplitude error $a_c(\omega)$ (which accounts for the power ratio α) in that range (see Fig. 6.3). Altering the gain in the path of the controller can result in an increase in cancellation of the primary noise for frequencies below 300 Hz. However, as Fig. 6.3 suggests, the cancellation for frequencies greater than 300 Hz may deteriorate and hence worsen the overall cancellation. Nevertheless, this experiment demonstrates that, by incorporating a model of the secondary source (loudspeaker) and the acoustic transmission paths in the design of the controller, noise cancellation over a broad frequency range of interest can be achieved. The level of cancellation achieved depends on how closely the filter represents the required controller characteristics and on the computational and processing power of the digital signal processor.

6.3.2 Practical source of noise

The exhaust noise of a motorcycle represents a typical example of a practical source of noise. It is a loud, low-frequency, and compact source of noise. The noise spectrum is essentially low frequency and is harmonically related to the fundamental (engine firing) frequency. The level of noise depends on the throttle setting; the noise is louder for increasing engine speeds. Such a characteristic is common to the noise of almost all machines of rotary type such as motor vehicles, electric machines, and fans.

With the primary source in Fig. 6.5 representing the exhaust noise of the motorcycle the cancellation as a function of frequency under idle engine speed will typically be as in Fig. 6.7. Here, noise is cancelled over a broad frequency range of about 62 Hz to about 373 Hz. The overall average amount of cancellation achieved is more than 4 dB with a maximum peak cancellation of 19.9 dB occurring at 284 Hz. However, the noise is reinforced at frequencies below 62 Hz and above 373 Hz. Moreover, the trend of cancellation as a function of frequency achieved is somewhat different from that in Fig. 6.6. This is due to the difference in locations of the detector and/or observer relative to the primary and secondary sources as compared with those used in obtaining Fig. 6.5, and further with those used in measuring the

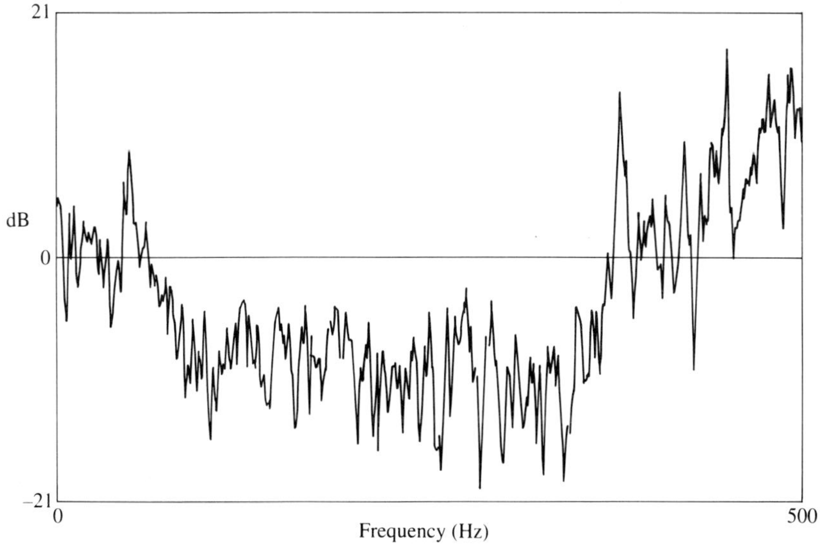

Fig. 6.7. Noise cancellation of the practical source

required controller characteristics (Fig. 6.1). This is one of the problems arising from such a design. The design relies on the assumption that the system geometry is the same under measurement of the controller transfer function and system operation. However, in practice this assumption can only hold to within certain limits. The controller here has a fixed transfer function that corresponds only to the arrangement in Fig. 6.1 under which the system will achieve good cancellation at observation points that are equidistant from the two sources. At low frequencies the effect of an error in the location of the detector and/or observer will not be significant in that the additional phase shift introduced in this process will be negligible. However, as the frequency increases this additional phase shift becomes more significant so that it eventually produces reinforcement. This practical problem, as well as the time-varying nature of many practical sources of noise, will require a controller that is capable of adapting its characteristics in accordance with the changes in the system so that a desired amount of cancellation is achieved over a broad frequency range.

6.4 Summary

The process of the design and implementation of a digital filter controller for use in an ANC system has been presented. This firstly involves the measurement of the transfer characteristics of the controller over a desired frequency range, secondly approximation of these characteristics by a suitable transfer function in the complex frequency s domain, thirdly approximation of this by an equivalent discrete-time transfer function via an appropriate s-plane to z-plane mapping, and finally implementation of the discrete-time transfer function on a dedicated digital signal processor.

The required controller for use in ANC systems has a transfer function that is dependent on a number of parameters in the system. Among these are the transfer characteristics and the geometrical set-up of the detector and observer microphones and the secondary source. A detector placed at a set distance relative to the primary and secondary sources of noise is intended to detect the frequency contents of the source noise without much alteration in the amplitude and phase of the wave. Moreover, the physical separation between the detector and/or observer relative to the sources should be such as to avoid situations of an infinite-gain controller requirement and problems of instability in the system.

In measuring the transfer function of the required controller the angle of orientation of an (omnidirectional) detector microphone with respect to the primary source has a direct effect on the resulting phase characteristic and consequently on the processes of design and implementation of the controller. An increase in the angle of orientation of the microphone relative to the primary source leads to an overall increase in phase characteristics of the required controller. Such an increase will require a transfer function of a higher order to approximate the measured characteristics. This in turn leads to increased computer data, processing time, and so on.

The characteristics of a secondary source loudspeaker have a similar effect on the controller transfer function. As the controller is intended to compensate for the transfer function of the acoustic and electronic components that are in the secondary source path (in addition to the acoustic paths between the observer and the sources) so that an exact mirror image of a detected noise is produced, the compensation for the secondary source loudspeaker will be of an inverting nature. This requires a loudspeaker with a flat amplitude characteristic and with least delay introduced due to its phase characteristic. Loudspeakers introduce large attenuations in signal amplitude at frequencies typically from zero to 50–100 Hz. This will require a controller to have large gain and low phase in this frequency range which in turn results in practical limitations in controller design and consequently unsuccessful operation of the ANC system at this range of frequencies.

Considering the practical limitations in the measurement, design, and implementation of the required controller transfer function an ANC system employing a digital controller performs to an impressive and significant level in the case of a compact, broadband source of noise. The performance can further be enhanced by using a powerful dedicated signal processor. However, the DFC, having a fixed transfer function and hence valid for a particular system geometry, will not perform well in the case of a change in the geometrical set-up of the system as well as under time-varying conditions. In these circumstances a system with the controller designed and implemented in an adaptive control framework is required.

6.5 References

Antoniou, A. (1979). *Digital filters: Analysis and design.* McGraw-Hill, New York.

Lam, H.Y-F. (1979). *Analog and digital filters: design and realisation.* Prentice-Hall, Englewood Cliffs.

Leitch, R.R. and Tokhi, M.O. (1987). Active noise control systems. *IEE Proceedings*, Part A, **134**, (6), 525–46.

Oppenheim, A.V., Willsky, A.S., and Young, I.T. (1984). *Signals and systems.* Prentice-Hall, Englewood Cliffs.

Temes, G.C. and Mitra, S.K. (1973). *Modern filter theory and design.* John Wiley, New York.

Terrel, T.J. (1980). *Introduction to digital filters.* Macmillan, London.

7

ADAPTIVE CONTROL

7.1 Introduction

The analysis, design, and implementation of ANC systems with compact sources of broadband noise in three dimensions under stationary conditions were considered in the previous chapters. The controller under such circumstances is thus realized with a fixed frequency-dependent transfer function. Assuming a linear propagation medium, such an ANC system results in significant amounts of noise cancellation if:

(1) the characteristics of the various components in the system, such as the source, transducers, and other electronic equipment used, are time-invariant; and
(2) the geometric set-up of the system is not altered once designed.

In practice, however, all the above conditions are not always met.

The characteristics of many practical sources drift and vary due to a number of factors, such as operating conditions. This leads to time-varying spectral characteristics. The characteristics of many transducers and other electronic equipment used are also subject to variations due to temperature, aging, and so on. The geometric set-up of the system may change due to transportation, for example. These factors were shown in the previous chapters to affect system stability and performance. Therefore, several practical situations exist under which an ANC system with a fixed controller is not adequate to achieve significant amounts of noise cancellation. Furthermore, the use of such a controller under time-varying conditions may result in reinforcement rather than cancellation. In such circumstances an alternative is to design an adaptive ANC system so that the controller characteristics are updated in accordance with variations in the system.

Adaptive systems are designed to modify their behaviour in accordance with the changing properties of controlled processes and their signals. An adaptive mechanism is characterized by the two complementary processes of identification and control. In the process of identification a suitable model is developed that exhibits the same input/output characteristics as the controlled process (plant). In the process of control a control action is determined, implemented, and tested on the plant on the basis of the identified model and control/performance objective. Among the various types of adaptive control algorithms, the most popular, which have proved successful in many applications, are the self-tuning adaptive controllers.

Self-tuning is a direct integration of an identification and a controller design algorithm in such a manner that the two processes proceed sequentially. Depending on the type of identification algorithm and controller design strategy employed various types of self-tuning algorithms can be designed. Many different identification schemes have been used in self-tuning control algorithms. Among these the recursive form of the least squares algorithm has proved to be the most useful and practically successful self-tuning identifier. The self-tuning controller was originally developed for the stochastic minimum variance control problem. However, since the approach is very flexible with respect to the controller design strategy different extensions are possible.

Self-tuning is, in a sense, the simplest possible adaptive control algorithm derivable from the point of view of discrete-time stochastic control theory. An attractive property of a self-tuning controller is that under most reasonable circumstances, as the number of input and output samples tends to infinity, it will converge to the optimal controller that would be obtained if the system parameters were exactly known. Moreover, the strategy is simple enough to allow the use of microprocessors for implementation, thus promising a relatively low-cost solution to complex control problems.

In this chapter the background information and general theory of adaptive systems are presented, leading to the introduction of self-tuning adaptive control and an introduction to system identification, concentrating specifically on the derivation and implementation aspects of the recursive least squares algorithm as a self-tuning identifier.

7.2 An overview of adaptive control systems

Adaptive control systems are designed to adapt their behaviour to the changing properties of controlled processes and their signals. Adaptive control has mainly been concerned with feedback systems having the classical structure shown in Fig. 7.1 where the controller performs two functions: it

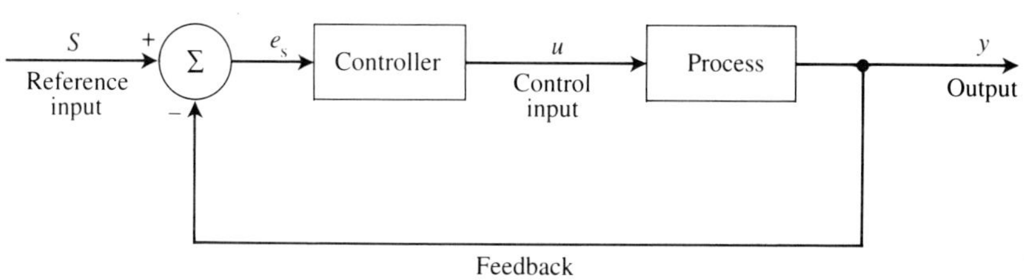

Fig. 7.1. Feedback control system

learns about the controlled process whilst, at the same time, controlling its behaviour.

The idea of adaptive control has its origins in the early days of control systems. However, it was not until the 1950s that serious attempts, due primarily to aerospace problems, were made to design practical adaptive systems. It was found that classical linear controllers in the configuration of Fig. 7.1 did not always give satisfactory control of the attitude of high-performance aircraft and rockets. The difficulty arose because response characteristics of these controlled processes varied significantly during flight, and the classical controller could be matched only to a single flight condition; under other conditions within the flight envelope the fixed classical controller would be unsatisfactory. There seemed to be a need for a more sophisticated controller which could automatically adapt itself to the changing characteristics of the controlled process. It was recognized that such a controller would have many other valuable applications, for example in industrial process control or in controlling economic systems, and that it might be analogous to some biological control systems.

Research on adaptive control was very active in the early 1950s. The work on adaptive flight control was characterized by a lot of enthusiasm, bad hardware, and weak theory (Mishkin and Braun 1961). Interest in the area diminished due primarily to two principal difficulties: lack of suitable computer technology for implementation and absence of adequate supporting theory. The contributions to control theory in the 1960s proved important for the development of adaptive control. State space and stability theory, results in stochastic control theory, dynamic programming, and dual control theory enhanced the understanding of adaptive processes. Fundamental contributions were made in recursive stochastic approximation and in system identification and parameter estimation (Astrom 1970, 1983*a*; Astrom and Eykhoff 1971; Isermann 1985; Wittenmark 1975). The interest in adaptive control was renewed in the 1970s. With the rapid and revolutionary progress in microelectronics and improved understanding of adaptive control it became feasible to experiment with various strategies for adaptive control. The methods ranged from simple gain adjustment to sophisticated algorithms which attempted to achieve optimal regulation in a stochastic environment. The field is now well established by researchers both at universities and in industry.

A model-based adaptive mechanism is characterized by two complementary phases of identification and control. In the identification phase, after making assumptions about the type of model, the parameters of the model are obtained using a system identification technique. In the control phase, given the structure and constraints imposed by the model, a control action is generated based on an implicit control strategy, implemented and tested on the plant. The two phases result in an adaptive controller that modifies its

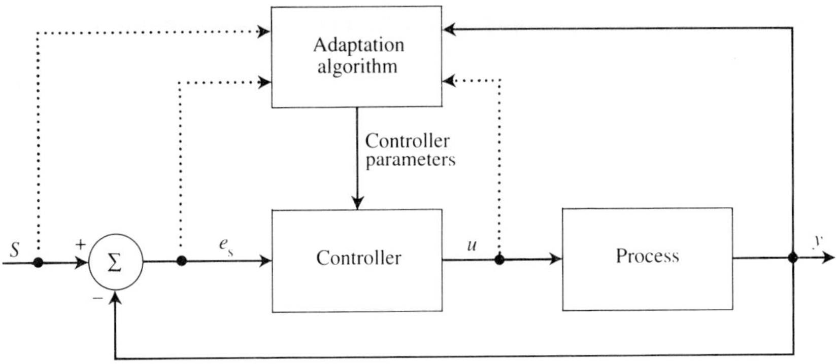

Fig. 7.2. Adaptation mechanism

response to changing characteristics of the plant. A supervisory level, based on a performance criterion, is often required so that a desired level of performance is maintained.

Two basic schemes of controller adaptation are shown in Fig. 7.2. Information on the process behaviour, such as structure and parameters, is obtained through a measurement of process input and output signals. This can be achieved, for instance, by process identification (measuring u and y) or by determining the closed-loop performance (measuring e_s and u). Based on such information the controller can be calculated and the control action modified.

Adaptive controllers can be divided into two main groups: model-reference and self-tuning. The general structure of model-reference adaptive controllers is shown in Fig. 7.3. Here, a control action is generated on the basis of the

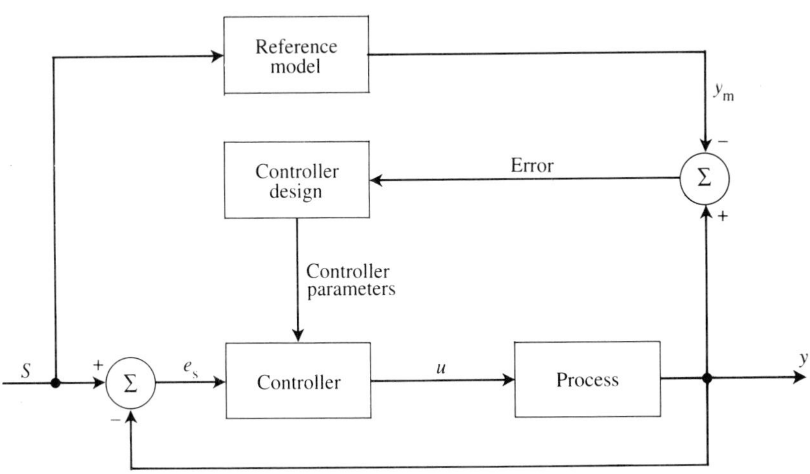

Fig. 7.3. Model-reference adaptive control

difference between the required system response (output of the reference model y_m) and the actual system response (output of the process y). The advantages of these controllers are quick adaptation for defined inputs (such as servo-control systems) and straightforward treatment of stability using non-linear stability theories.

The general structure of self-tuning adaptive controllers is shown in Fig. 7.4. Here, the desired performance given by the control objective is reached on the basis of a given controller type and the obtainable information on the process and its signals. Self-tuning adaptive controllers have the advantage of being able to adapt to any (especially unmeasurable) disturbances, if properly designed.

Self-tuning algorithms and model-reference controllers share the same basic structure. Their control input u is the certainty-equivalent control for an assumed linear controlled process with uncertain coefficients, such as

$$\sum_{i=0}^{ma} a_i y(nT - iT) = \sum_{i=0}^{mb} b_i u(nT - iT) + \sum_{i=0}^{mf} f_i g(nT - iT) + \zeta(nT) \quad (7.1)$$

where the coefficients a and b are the process parameters, the coefficients f are to take into account the effect of the unobservable white-noise signal g, the signals u and y, respectively, are the system input and output, ζ represents the bias due to measurable extraneous signals, T is the sampling period, and n is an integer. Note that some of the leading coefficients b in equation (7.1) may be zero to account for the delay between u and y.

In self-tuning and model-reference controls the control signal u is generated by a feedback control law having adjustable coefficients which are tuned

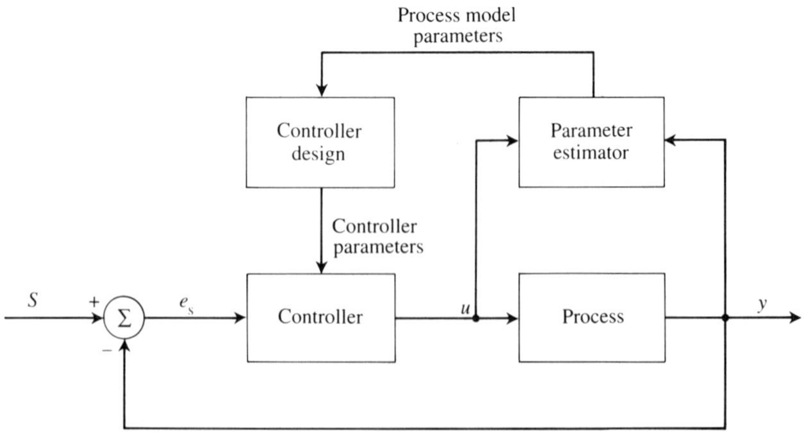

Fig. 7.4. Self-tuning control

or adapted by the algorithm. These two classes were originally derived from very different points of view, but have now come to be recognized as closely related, especially for the controller itself and the convergence analysis (Astrom and Wittenmark 1989; Egardt 1980; Gawthrop 1977; Isermann 1982; Landau 1979; Narendra and Lin 1980).

7.2.1 *Self-tuning control*

Self-tuning control is, in a sense, the simplest possible adaptive control algorithm derivable from the point of view of discrete-time stochastic control theory. The identification algorithm is simplified by making the assumption that all the coefficients f in equation (7.1) are zero, except for f_0 which has the value unity, that is

$$F(z) = 1. \qquad (7.2)$$

A further simplification is achieved by neglecting the bias ζ in equation (7.1). In a more general case a suitable model is obtained for the measurable extraneous signals. This is discussed later under system identification. The effect of the assumption in equation (7.2) is to eliminate uncertainty about the coefficients f so that the problem of estimating the remaining coefficients is linearized. This linearization is demonstrated by rewriting equation (7.1) using a vector θ to represent the remaining uncertain coefficients

$$\theta^T = [\,-a_1, \ldots, -a_{ma}, b_0, \ldots, b_{mb}] \qquad (7.3)$$

together with a vector ψ to represent the relevant current and past input sequences and the past output sequence affecting the current output $y(nT)$:

$$\psi = [y(nT - T), \ldots, y(nT - maT), u(nT), \ldots, u(nT - mbT)]. \qquad (7.4)$$

Then, using the notations of equations (7.3) and (7.4) together with the assumption of equation (7.2) and neglecting ζ, the assumed process of equation (7.1) can be written as

$$y(nT) = \psi\theta + g(nT). \qquad (7.5)$$

This equation is clearly linear in the uncertain coefficients θ. It is a starting point for classical estimation theory which leads to least squares estimation algorithms that generate an unbiased minimal-variance estimate θ and can conveniently be implemented recursively and in real time.

Self-tuning control assumes that the controlled process satisfies equation (7.5). It combines the simple recursive least squares estimation of coefficients θ with a similarly simple certainty-equivalent control law having parameters which are adjusted using the resulting estimates $\hat{\theta}$.

Self-tuning adaptive controllers can be classified according to the following.

Process model The implementation on a digital computer and the availability of proper controller design procedures lead to discrete-time process models in parametric form. Therefore difference equations such as equation (7.1) are mostly used as process models.

Process identification methods For the identification of parametric process models the use of parameter estimation methods is straightforward. The recursive form of parameter estimation methods are best suited for on-line identification in real time.

Information about the process Information about the process thus consists of the parameter estimates $\hat{\theta}$ and, if desired, of their uncertainty $\Delta\hat{\theta}$.

If the parameter estimates $\hat{\theta}$ are assumed to be identical with the process parameters θ, the resulting controllers are called certainty-equivalence controllers. However, if in addition to $\hat{\theta}$ the uncertainties $\Delta\hat{\theta}$ of the parameter estimates are also taken into account for the controller calculation, the resulting controllers are called cautious controllers (Wittenmark 1975).

Controller design criterion The performance of adaptive control systems, based on process identification, mainly depends on the quality of the identified process model and therefore on the process input signals. Hence, the input signal has to be determined so that a good present control as well as a good future process identification is achieved. The inclusion of both requirements in the controller design criterion leads to dual controllers. Non-dual controllers use only the present and past information about the process in their design criteria (Wittenmark 1975). Frequently used design criteria for non-dual controllers are quadratic in the system input and output or are special design criteria, for example, minimum variance, principle of deadbeat, pole–zero cancellation, or pole-assignment (Astrom and Wittenmark 1973, 1980; Clarke and H-James 1971; Clarke and Gawthrop 1979; Wellstead and Zanker 1979; Wellstead *et al.* 1979*a*,*b*; Wittenmark 1974; Yamane *et al.* 1983).

Control algorithm The actual design of the control algorithm is, of course, done before its implementation in the digital computer. Control algorithms are designed on the basis of satisfying:

(1) closed-loop identifiability conditions;
(2) small computational expense and storage requirements for the controller parameter calculation; and
(3) applicability for many classes of processes and signals.

Within the class of self-tuning adaptive controllers those types which have demonstrated most success, both in practice and theory, are characterized by recursive parameter estimation, the non-dual, and the certainty-equivalence principle. This kind is also called the self-tuning regulator (Astrom 1983*a*; Astrom and Wittenmark 1973; Astrom *et al.* 1977; Wellstead *et al.* 1979*a,b*; Wittenmark 1974).

A self-tuning adaptive control mechanism (Fig. 7.4) can be thought of as composed of two loops. An inner loop consisting of the process and an ordinary linear feedback controller and an outer loop composed of a recursive parameter estimator and a controller design calculation. The parameters of the linear feedback controller are thus adjusted by the outer loop.

Notice that the box labelled 'controller design' in Fig. 7.4 represents an on-line solution to a design problem for a system with known parameters. This is called the underlying design problem. Such a problem can be associated with most adaptive control schemes. However, the problem is often given indirectly. To evaluate adaptive control schemes it is often useful to find the underlying design problem because it will give the characteristics of the system under the ideal conditions when the parameters are known.

Self-tuning was originally proposed, under the title 'self-optimizing', by Kalman (1958) who described a mid-1950s hybrid computer implementation using multi-turn potentiometers and thermionic valves. The proposal had little immediate consequence because of two difficulties:

(1) computing technology was not sufficiently advanced;
(2) equation (7.5) seemed to be too specialized to represent any very general class of practical controlled processes.

Both these problems were solved subsequently thus making self-tuning a very popular development area in adaptive control. Difficulty (1) has been resolved due to the widespread availability of mini- and microcomputers. The breakdown of difficulty (2) has primarily been due to valuable contributions of Astrom and his co-workers.

The self-tuning controller was originally developed for the stochastic minimum variance control problem (Astrom and Wittenmark 1973). Since the approach is very flexible with respect to the underlying design method many different extensions have been made, for example, extended minimum variance design, pole-assignment design, self-tuners based on the LQG (Linear Quadratic Gaussian) design method and on phase and amplitude margins (Astrom 1983*b*; Astrom and Hagglund 1983, 1984; Astrom and Wittenmark 1973, 1980; Astrom and Zhao-Ying 1982; Bohm 1985; Clarke and Gawthrop 1975, 1979; Gawthrop 1980; Harris and Billings 1981; Nagy and Peterka 1985; Peterka 1970; Qin 1985; Xue-Shan 1985; Wellstead and Sanoff 1981; Wellstead and Zanker 1979; Wellstead *et al.* 1979*a,b*).

The self-tuner also contains a recursive parameter estimator. Many different estimation schemes have been used, for example, stochastic approximation, least squares, extended and generalized least squares, instrumental variables, extended Kalman filtering, and the maximum likelihood method (Astrom 1983*a*; Astrom and Eykhoff 1971; Isermann 1982; Harris and Billings 1981; Ljung and Soderstrom 1983; Strejc 1980).

The philosophy of self-tuning control is to adjust the controller parameters on the basis of current control behaviour in order that future behaviour may be improved. One attractive property of a self-tuning controller is that under most reasonable circumstances, as the number of input and output samples tends to infinity, it will converge to the optimal controller that would be obtained if the system parameters were exactly known. This is known as the self-tuning property which makes the application of self-tuning feasible for solving practical problems (Astrom and Zhao-Ying 1982; Clarke and Gawthrop 1979; Wellstead *et al.* 1979*b*; Wittenmark 1974). Furthermore, the strategy is simple enough to allow the use of microprocessors for implementing self-tuning, thus providing a relatively low-cost solution to complex control problems.

In essence self-tuning is the direct combination of an identification and a controller design algorithm so that the two processes proceed one after the other. Self-tuning control algorithms can be divided into two categories: explicit and implicit. In an explicit (or indirect) algorithm the process model parameters are estimated first, stored as an intermediate result, and then used with a design strategy to calculate the controller parameters (Astrom 1983*a*; Astrom and Wittenmark 1980; Kurz *et al.* 1980). In an implicit (or direct) algorithm the parameters of the controller are estimated directly (Astrom 1983*a*; Astrom and Wittenmark 1973, 1980; Clarke and Gawthrop 1975; Peterka 1970; Xue-Shan 1985). An implicit algorithm can be designed by a re-parametrization of the process model, that is, by introducing the controller design equation into the process model equation, so that the model parameters are not estimated and the estimator gives the controller parameters directly.

An advantage of the implicit algorithms over the explicit algorithms is that the design computations are eliminated, since the controller parameters are estimated directly. Moreover, simulations and practical experiments have shown that implicit algorithms are more robust (Kosut 1983; Praly 1983). However, implicit algorithms usually have more parameters to estimate than their explicit counterparts, especially if there are long delays in the process. A further disadvantage is that with implicit algorithms all process zeros are cancelled. This implies that implicit methods are intended only for processes with a stable inverse (minimum-phase systems). However, through a proper choice of parameters, many implicit algorithms can also be used if the system is non-minimum phase (Astrom and Wittenmark 1985).

The problems of convergence analysis on system identification are extensively discussed in the literature (Eykhoff 1981; Ljung and Soderstrom 1983;

Strejc 1979). However, for an adaptive controller which is actually nothing more than a special non-linear control algorithm, motivated by combining on-line parameter estimation with on-line control, the essential non-linearity of the algorithm has proven to be a major stumbling block in establishing the necessary convergence properties. There are two complications in this case.

1. Since the process input is generated by feedback the excitation of the process depends on the process disturbances. It is then difficult to show that the input is persistently exciting (a necessity for convergence of the estimate).
2. The input is also correlated with the disturbances because it is generated by feedback.

Another issue of current concern in adaptive control is the robustness of the algorithms. For an algorithm to perform well, an accurate controller design procedure on the basis of a good process model is required. This requirement leads to the question of how to obtain a good model. The key issues are good data and an appropriate model structure.

Research in adaptive control theory has focused almost exclusively on the case where the plant can be fully represented by some member of a family of linear finite-dimensional parametric models (Goodwin *et al.* 1980; Narendra *et al.* 1980). Thus, the model error due to unmodelled dynamics can cause adaptive controllers to exhibit significant performance degradation and instability, even with an initial controller parametrization that closely approximates the desired closed-loop response (Ionnou 1983; Kosut 1983; Rohrs *et al.* 1982). These simulated circumstances of undesirable behaviour are in sharp contrast to successful applications of adaptive control where reduced-order modelling is unavoidable (Astrom 1983*a*). This issue of model error is of undeniable practical importance because no actual plant is truly linear or finite-dimensional. To estimate an accurate reduced-order model it is essential that the input signal has sufficient energy content and is rich in frequency that it is persistently exciting. A necessary condition to obtain a robust adaptive control algorithm is to use both robust control and robust estimation.

7.3 System identification

People have been concerned with interpreting and making estimates and predictions since earliest times. Galileo Galilei is credited with initiating a theory of estimation in 1632 attempting to minimize various functions of errors. However, it was Gauss who first proposed, in 1795, the method of least squares as a means of predicting the motion of planets based on astronomical observations. Since then the method of least squares has been applied for the solution of many technical problems. Its properties have been analysed many times, numerical procedures have been proposed for individual applications, and the method itself has been modified according to specific requirements. The endeavour may be observed in the field of parameter estimation of controlled systems. The techniques developed for input/output data collection and evaluation to solve the problem of identifying a dynamic process range from the simplest form of the deterministic procedures to elegant numerical and statistical methods based on results of optimal estimation theory.

The identification problem is frequently formulated as: given a class of models, a criterion, and measurements of the input and output signals, find the particular model which fits the experimental data best in the sense of the given criterion. Various methods for solving the identification problem differ in the choice of models and criteria as well as mathematical techniques.

Process identification can be performed off-line or on-line. For off-line identification the process data is first stored in a data storage medium. The data is later transferred to a computer and evaluated. Batch processing of the data is mostly applied to this type of identification, that is, the whole data set is evaluated once.

In on-line identification the process identification is performed during on-line operation. Two ways of processing the data can be distinguished: real-time processing, where the data is evaluated immediately after each sample, and batch processing, where the data set is evaluated only after all the measurements have been made.

Real-time processing in general uses less storage of data; the model develops as the data comes in. Recursive identification algorithms are normally used in real-time processing. These algorithms allow the identification of time-varying processes. In batch processing the data is stored and evaluated at the end of the measurements. Here, non-recursive identification algorithms are mostly used.

In self-tuning control the parameters of a system model are obtained by first assuming a structure for the system model, and then using an estimation algorithm to obtain estimates of the parameters of the model. This requires putting the system and signal representations, for a computer control scheme, into a form suitable for parameter estimation, and then using the recursive form of the least squares algorithm.

7.3.1 The process model

A process can generally be represented, in the discrete-time domain, by a linear combination of the system (plant) model and a disturbance model; thus:

$$y(z) = \frac{B(z)}{A(z)} u(z) + \xi(z) \tag{7.6}$$

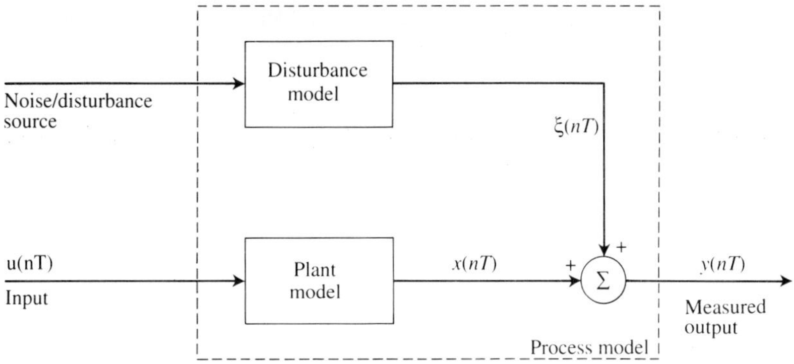

Fig. 7.5. Process model

where u, y, and ξ are, respectively, the input, measured output, and disturbance signals and $A(z)$ and $B(z)$ are polynomials in the z-transform variable z:

$$\begin{aligned} A(z) &= 1 + a_1 z^{-1} + a_2 z^{-2} + \cdots + a_{ma} z^{-ma} \\ B(z) &= b_0 + b_1 z^{-1} + b_2 z^{-2} + \cdots + b_{mb} z^{-mb}. \end{aligned} \tag{7.7}$$

This is shown in Fig. 7.5 where $x(nT)$ is the output due to $u(nT)$, T is the sampling time, and n is an integer. Note that some of the leading coefficients in $B(z)$ may be zero due to a constant delay between $u(nT)$ and $y(nT)$. Equation (7.6) is identical with the z-transform of equation (7.1) with ξ representing a general disturbance signal including the white-noise term q and the bias term ζ. Modelling of the disturbance requires a knowledge of the extraneous signals affecting the measured output. Therefore, a brief description of the disturbance signals is given.

The disturbance signals can broadly be classified as either deterministic or stochastic. In the deterministic case ξ can take one (or a combination) of the following forms.

Offset A constant value may shift the system output, $x(nT)$ in Fig. 7.5, by a constant amount.

Drift A generalization of the constant offset is output drift whereby the offset becomes a function of time. In many situations drift can be modelled by a polynomial function of time. A further form of time-dependent drift occurs when the offset is periodic. Periodic drift can be represented in a model by the appropriate periodic function. Although time dependency is usual, any independent variable that affects the offset can be used.

Measurable disturbance Often the system output $x(nT)$ is corrupted by a disturbance from a known source, influencing the output via a disturbance transfer function. For convenience the disturbance transfer function is usually assumed to have the same denominator as the system transfer function, so that if v represents the signal from the known source then the disturbance model is

$$\xi(z) = \frac{H(z)}{A(z)} v(z) \tag{7.8}$$

where $H(z) = h_0 + h_1 z^{-1} + \cdots + h_{mh} z^{-mh}$.

Note in equation (7.8) that some of the leading coefficients in $H(z)$ may be zero, corresponding to a time delay in the disturbance. Furthermore, unless the signal v is z-transformable the representation in equation (7.8) will be incorrect. In practice, the errors caused are small compared with the improved control performance obtained by modelling the disturbance.

The most important forms of stochastic or random disturbances are those associated with small unpredictable changes in a system and unobservable noise-like disturbances. Such disturbances are aggregated and modelled by a noise source which can be assumed to be stationary; a Gaussian noise sequence. This can be represented by a white Gaussian noise sequence $g(nT)$, with mean zero and variance σ_g^2, passed through a linear time-invariant filter N (Fig. 7.6). If $\xi(nT)$ is the noise sequence as measured at the output by setting $u(nT) = 0$, and assuming no other disturbances are present, then $\xi(nT)$ will have a power spectrum given by

$$G_{nn}(\omega) = N(e^{j\omega})N(e^{-j\omega})\sigma_g^2. \tag{7.9}$$

From a measurement of $G_{nn}(\omega)$ a minimum phase and stable noise model N is constructed to have the measured power spectrum. In practice it is commonly assumed that the noise filter and system model share the same denominator; thus:

$$\xi(z) = \frac{F(z)}{A(z)} g(z) \tag{7.10}$$

where $F(z) = 1 + f_1 z^{-1} + f_2 z^{-2} + \cdots + f_{mf} z^{-mf}$. If the system model and noise filter do not have the same poles then common factors can be introduced in $A(z)$, $B(z)$, and $F(z)$ to make a common denominator.

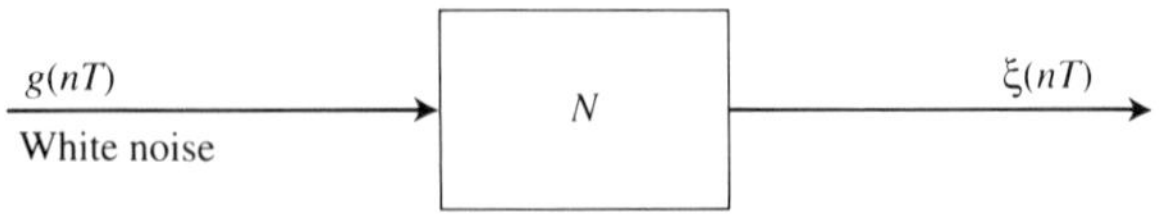

Fig. 7.6. Noise filter

7.3.2 *Linear least squares identification*

Consider a discrete-time transfer function model of a system with input sequence $u(nT)$ and output $y(nT)$ subject to disturbances from measurable source $v(nT)$, drift noise $w(nT)$, and random noise $g(nT)$ as shown in Fig. 7.7. Thus

$$A(z)y(z) = B(z)u(z) + H(z)v(z) + W(z) + F(z)g(z) \tag{7.11}$$

where

$$A(z) = 1 + a_1 z^{-1} + \cdots + a_{ma} z^{-ma}$$
$$B(z) = b_0 + b_1 z^{-1} + \cdots + b_{mb} z^{-mb}$$
$$H(z) = h_0 + h_1 z^{-1} + \cdots + h_{mh} z^{-mh}$$
$$F(z) = 1 + f_1 z^{-1} + \cdots + f_{mf} z^{-mf}$$
$$W(z) = w_0 + w_1 nT + \cdots + w_{mw}(nT)^{mw}.$$

When the system to be controlled is known then the coefficients of the polynomials $A(z)$, $B(z)$, $F(z)$, $H(z)$, and $w(nT)$ can be calculated from system parameters. However, if the system is unknown then the polynomial coefficients are treated as parameters to be measured or estimated. The disturbances from measurable source and drift noise can be treated as measurable, or known, variables. For the random noise sequence, since $g(nT - T)$, $g(nT - 2T), \ldots, g(nT - mfT)$ are the past values of the unobservable white-noise disturbance $g(nT)$ and in general will be unknown, it is assumed that $mf = 0$, that is, $f_1 = \ldots = f_{mf} = 0$ or $F(z) = 1$.

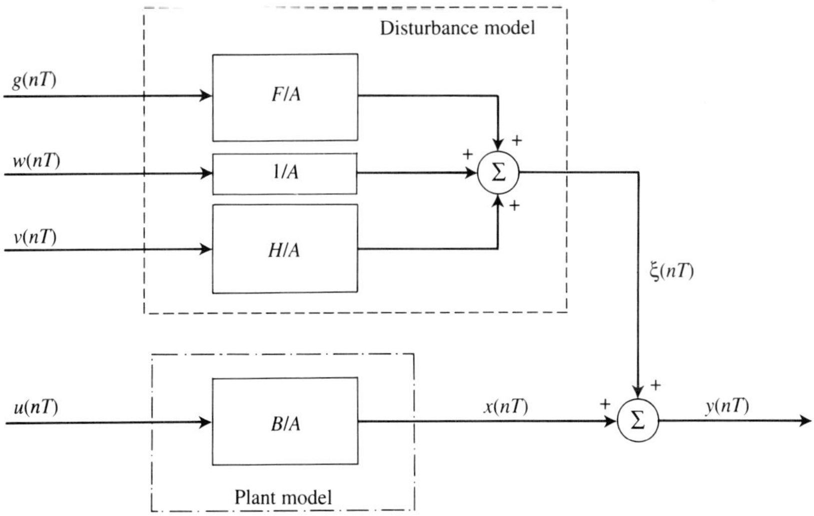

Fig. 7.7. Process model

Equation (7.11) can be written, in the discrete-time domain, in matrix form as

$$y(nT) = \psi\theta + g(nT) \tag{7.12}$$

where θ is the vector of unknown parameters defined by

$$\theta^{\mathrm{T}} = [-a_1, \ldots, -a_{ma}, b_0, \ldots, b_{mb}, h_0, \ldots, h_{mh}, w_0, \ldots, w_{mw}] \tag{7.13}$$

and ψ, known as the observation matrix, is a row vector of measured input/output variables:

$$\begin{aligned}\psi = [&y(nT-T), \ldots, y(nT-maT), u(nT), \ldots, u(nT-mbT),\\ &v(nT), \ldots, v(nT-mhT), 1, nT, \ldots, (nT)^{mw}].\end{aligned} \tag{7.14}$$

Assume that the process described by equation (7.12) has been running for sufficient time to form k sets of the data vector ψ, that is, $\psi(1), \ldots, \psi(k)$. The data obtained in this way can be assembled into a matrix form to enable us to solve for the unknown parameters in the vector θ. Thus

$$\begin{bmatrix} y_1(nT) \\ y_2(nT) \\ \vdots \\ y_k(nT) \end{bmatrix} = \begin{bmatrix} \psi(1) \\ \psi(2) \\ \vdots \\ \psi(k) \end{bmatrix} \theta + \begin{bmatrix} g_1(nT) \\ g_2(nT) \\ \vdots \\ g_k(nT) \end{bmatrix}. \tag{7.15}$$

This is the specification of the true system in which θ is a vector of true system parameters and $[g_1(nT), g_2(nT), \ldots, g_k(nT)]^{\mathrm{T}}$ is the unobservable, uncorrelated white noise having the following properties (Brown 1983).

$$\begin{aligned}
E\{g(qT)\} &= 0, & & q = 1, 2, \ldots, k\\
E\{g(qT)g(qT-\tau T)\} &= 0, & \tau &\neq 0, & q = 1, 2, \ldots, k\\
E\{g(qT)u(qT-\tau T)\} &= 0, & \tau &= 0, 1, \ldots, k, & q = 1, 2, \ldots, k\\
E\{g(qT)y(qT-\tau T)\} &= 0, & \tau &= 0, 1, \ldots, k, & q = 1, 2, \ldots, k\\
E\{g(qT)g(qT)\} &= \sigma_{\mathrm{g}}^2, & & q = 1, 2, \ldots, k
\end{aligned} \tag{7.16}$$

where $E\{.\}$ designates the expected value and σ_{g}^2 is the variance of $g(nT)$.

To determine estimates of the system parameters a model for the system is assumed as

$$\begin{bmatrix} y_1(nT) \\ y_2(nT) \\ \vdots \\ y_k(nT) \end{bmatrix} = \begin{bmatrix} \psi(1) \\ \psi(2) \\ \vdots \\ \psi(k) \end{bmatrix} \hat{\theta} + \begin{bmatrix} e(1) \\ e(2) \\ \vdots \\ e(k) \end{bmatrix}. \tag{7.17}$$

where $\hat{\theta}$ is a vector of parameters for a system model and $e(i)$ is the modelling error at time step i. The aim is to select $\hat{\theta}$ so that the modelling error is minimized in some way (see Fig. 7.8). In some cases this will mean that the

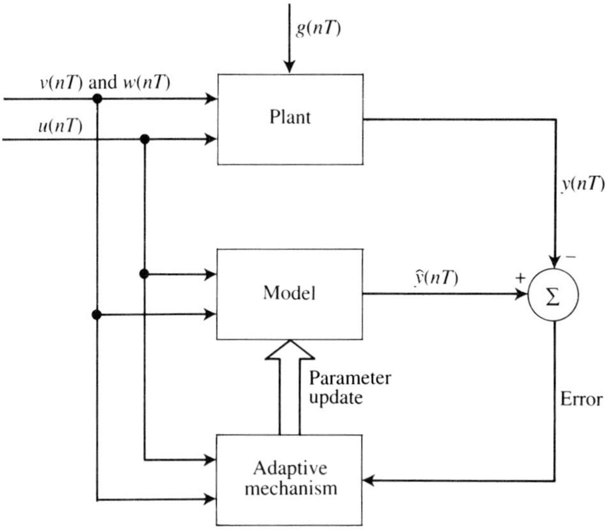

Fig. 7.8. Process identification

modelling errors (also called the fitting error or residuals) will become equal to the white noise $g(nT)$ disturbing the system.

To estimate the parameters uniquely the number of equations in equation (7.17) must exceed the number of parameters in the vector $\hat{\theta}$. In the noise-free case, that is, when $g(nT) = 0$, equation (7.17) can be solved as a set of linear equations in $m = ma + mb + mh + mw + 3$ unknowns. When noise is present (and in practice, even in nominally noise-free systems) the number of equations must be much larger than m and some statistical procedure to reduce estimation errors induced by the noise should be used. The technique most widely used in this connection is linear least squares which is derived as follows.

Rewriting equation (7.17) in matrix form gives

$$Y_k = \Psi_k \hat{\theta} + E_k \tag{7.18}$$

where

$$\begin{aligned} Y_k^{\mathrm{T}} &= [y_1(nT), \ldots, y_k(nT)] \\ E_k^{\mathrm{T}} &= [e(1), \ldots, e(k)] \\ \Psi_k &= \begin{bmatrix} \psi(1) \\ \psi(2) \\ \vdots \\ \psi(k) \end{bmatrix}. \end{aligned} \tag{7.19}$$

Rearranging equation (7.18) yields

$$E_k = Y_k - \Psi_k\hat{\theta}. \tag{7.20}$$

Define a cost function J as the sum of the squares of errors:

$$J = \sum_{i=1}^{k} e^2(i) \tag{7.21}$$

or

$$J = E_k^{\mathrm{T}} E_k. \tag{7.22}$$

In linear least squares the parameter vector $\hat{\theta}$ is chosen such that the cost function J is a minimum.

Substituting for E_k from equation (7.20) into equation (7.22) yields

$$J = (Y_k - \Psi_k\hat{\theta})^{\mathrm{T}}(Y_k - \Psi_k\hat{\theta})$$

or

$$J = Y_k^{\mathrm{T}} Y_k - \hat{\theta}^{\mathrm{T}}\Psi_k^{\mathrm{T}} Y_k - Y_k^{\mathrm{T}}\Psi_k\hat{\theta} + \hat{\theta}^{\mathrm{T}}\Psi_k^{\mathrm{T}}\Psi_k\hat{\theta}.$$

Differentiating this with respect to $\hat{\theta}$ and simplifying gives

$$\begin{aligned} \frac{\partial J}{\partial \hat{\theta}} &= -2\Psi_k^{\mathrm{T}} Y_k + 2\Psi_k^{\mathrm{T}}\Psi_k\hat{\theta} \\ \frac{\partial^2 J}{\partial \hat{\theta}^2} &= 2\Psi_k^{\mathrm{T}}\Psi_k. \end{aligned} \tag{7.23}$$

It follows from equations (7.19) and (7.23) that $\partial^2 J/\partial\hat{\theta}^2$ is non-negative. Therefore, J is a minimum for a vector $\hat{\theta}$ for which $\partial J/\partial\hat{\theta}$ is equal to zero. Equating $\partial J/\partial\hat{\theta}$ in equation (7.23) to zero and simplifying yields the parameter estimates $\hat{\theta}$ as

$$\hat{\theta} = [\Psi_k^{\mathrm{T}}\Psi_k]^{-1}\Psi_k^{\mathrm{T}} Y_k. \tag{7.24}$$

This is the linear least squares estimator of the system parameter vector θ.

The least squares estimates $\hat{\theta}$ can now be related to equation (7.15) which defines the plant and disturbances. Substituting for Y_k from equation (7.15) into equation (7.24) yields

$$\hat{\theta} = [\Psi_k^{\mathrm{T}}\Psi_k]^{-1}[\Psi_k^{\mathrm{T}}\Psi_k\theta + \Psi_k^{\mathrm{T}} G_k]$$

or

$$\hat{\theta} = \theta + [\Psi_k^{\mathrm{T}}\Psi_k]^{-1}\Psi_k^{\mathrm{T}} G_k \tag{7.25}$$

where $G_k^{\mathrm{T}} = [g_1(nT), \ldots, g_k(nT)]$. The deviation of the estimate of the parameter vector from its true value (bias in the estimator) is thus

$$\hat{\theta} - \theta = [\Psi_k^{\mathrm{T}}\Psi_k]^{-1}\Psi_k^{\mathrm{T}} G_k. \tag{7.26}$$

It follows from the properties of the white noise $g(nT)$ in equation (7.16) that the expected value of the parameter error vector $\hat{\theta} - \theta$ is

$$E\{\hat{\theta} - \theta\} = E\{[\Psi_k^{\mathrm{T}}\Psi_k]^{-1}\Psi_k^{\mathrm{T}}\}E\{G_k\}$$

or

$$E\{\hat{\theta} - \theta\} = 0.$$

Therefore, in the mean sense, the least squares estimate of the parameter error is zero. In other words, the parameter estimates are unbiased.

The covariance of the estimates is defined by

$$\mathrm{cov}(\hat{\theta}) = E\{\hat{\theta} - \theta)(\hat{\theta} - \theta)^{\mathrm{T}}\}.$$

Substituting for $\hat{\theta} - \theta$ from equation (7.26) yields

$$\mathrm{cov}(\hat{\theta}) = E\{([\Psi_k^{\mathrm{T}}\Psi_k]^{-1}\Psi_k^{\mathrm{T}}G_k)([\Psi_k^{\mathrm{T}}\Psi_k]^{-1}\Psi_k^{\mathrm{T}}G_k)^{\mathrm{T}}\}$$

or

$$\mathrm{cov}(\hat{\theta}) = [\Psi_k^{\mathrm{T}}\Psi_k]^{-1}\Psi_k^{\mathrm{T}}E\{G_kG_k^{\mathrm{T}}\}\Psi_k[\Psi_k^{\mathrm{T}}\Psi_k]^{-1}. \tag{7.27}$$

It follows from equation (7.16) that

$$E\{G_kG_k^{\mathrm{T}}\} = \sigma_g^2 I \tag{7.28}$$

where I is the unit matrix. Substituting for $E\{G_kG_k^{\mathrm{T}}\}$ from equation (7.28) into equation (7.27) and simplifying yields the covariance of the errors in the least squares parameter estimates as

$$\mathrm{cov}(\hat{\theta}) = \sigma_g^2[\Psi_k^{\mathrm{T}}\Psi_k]^{-1}. \tag{7.29}$$

It follows from equation (7.29) that the matrix $[\Psi_k^{\mathrm{T}}\Psi_k]^{-1}$, known as the covariance matrix, gives a direct measure of the variability and co-variability of the parameter estimates. Therefore, this matrix is of great importance in the identification process of self-tuning control. Note that:

1. The dimensions of the covariance matrix depend upon the number of unknown parameters, not the number of data samples. In general for m unknown parameters the covariance matrix is of order $m \times m$.
2. The entries in the covariance matrix are correlation-like sums over the available k data points and the matrix is symmetric.
3. If $u(nT)$ is a constant u_0, then

$$\Psi_k^{\mathrm{T}}\Psi_k = ku_0^2\begin{bmatrix} 1 & 1 & . & . & . & 1 \\ 1 & 1 & . & . & . & 1 \\ \vdots & \vdots & \vdots & \vdots & \vdots & \vdots \\ 1 & 1 & . & . & . & 1 \end{bmatrix}$$

which is singular and hence a unique least squares solution cannot be obtained because this matrix is non-invertible. This corresponds to a set of simultaneous equations where the number of equations is less than the number of unknowns. For a solution to be obtained $u(nT)$ must vary sufficiently to ensure that, as k increases,

$$\frac{1}{k}\det[\Psi_k^{\mathrm{T}}\Psi_k] \neq 0$$

where 'det' denotes 'determinant of'. Such conditions are usually associated with the term 'persistently exciting'. In self-tuning it is important that $u(nT)$ changes sufficiently to avoid a rank-deficient covariance matrix.

A brief description of the relation between the residuals or fitting error sequence E_k, obtained from least squares, and the data is given as follows. Pre-multiplying equation (7.18) by Ψ_k^{T} yields

$$\Psi_k^{\mathrm{T}} Y_k = [\Psi_k^{\mathrm{T}}\Psi_k]\hat{\theta} + \Psi_k^{\mathrm{T}} E_k.$$

Substituting for $\Psi_k^{\mathrm{T}} Y_k$ from the above into equation (7.24) yields

$$\hat{\theta} = [\Psi_k^{\mathrm{T}}\Psi_k]^{-1}([\Psi_k^{\mathrm{T}}\Psi_k]\hat{\theta} + \Psi_k^{\mathrm{T}} E_k).$$

Simplifying this gives

$$[\Psi_k^{\mathrm{T}}\Psi_k]^{-1}\Psi_k^{\mathrm{T}} E_k = 0$$

or

$$\Psi_k^{\mathrm{T}} E_k = 0. \tag{7.30}$$

Substituting for Ψ_k from equation (7.19) into equation (7.30) yields

$$[\psi^{\mathrm{T}}(1), \psi^{\mathrm{T}}(2), \ldots, \psi^{\mathrm{T}}(k)]E_k = 0.$$

This, using equation (7.14) and E_k from equation (7.19) yields

$$\begin{aligned} &\sum_{j=1}^{k} y_j(nT - iT)e(j) = 0; \qquad i = 1, \ldots, ma \\ &\sum_{j=1}^{k} u_j(nT - iT)e(j) = 0; \qquad i = 0, \ldots, mb \end{aligned} \tag{7.31}$$

and so on, for the remaining entries in Ψ. Assuming the process under consideration is ergodic, then as k becomes large the following will hold

$$\begin{aligned} &E\{y_j(nT - iT)e(j)\} = 0; \qquad j = 1, \ldots, k,\ i = 1, \ldots, ma \\ &E\{u_j(nT - iT)e(j)\} = 0; \qquad j = 1, \ldots, k,\ i = 0, \ldots, mb. \end{aligned} \tag{7.32}$$

Equations (7.31) and (7.32) are sometimes referred to as the orthogonality properties of least squares, and are the basis of a number of results, one being the self-tuning property.

7.3.3 *Recursive least squares identification*

Equation (7.24) is a least squares estimator of the parameter vector θ of the system model. This is used at each sample interval, as new input/output data becomes available, with a current model to generate a new fitting error. Instead of recalculating the least squares in its entirety it is preferable to update the previous estimates $\hat{\theta}(k-1)$ calculated at time $(k-1)T$ to obtain new estimates $\hat{\theta}(k)$ at time kT.

Assume that the system has been running for sufficient time to allow collection of the input/output samples from time steps 1 to k (that is, k samples). Thus, from equation (7.24) the parameter vector $\hat{\theta}(k)$ at time step k is

$$\hat{\theta}(k) = [\Psi_k^{\mathrm{T}}\Psi_k]^{-1}\Psi_k^{\mathrm{T}} Y_k \tag{7.33}$$

or

$$\hat{\theta}(k) = P(k)Q(k) \tag{7.34}$$

where Ψ_k and Y_k are data matrices from time steps 1 to k as given in equation (7.19) and the matrices $P(k)$ and $Q(k)$ at time step k are

$$\begin{aligned} P(k) &= [\Psi_k^{\mathrm{T}}\Psi_k]^{-1} \\ Q(k) &= \Psi_k^{\mathrm{T}} Y_k. \end{aligned} \tag{7.35}$$

Note that $P(k)$ is the covariance matrix at time step k, given in equation (7.29).

Substituting for Ψ_k from equation (7.19) into equation (7.35) yields the covariance matrix as

$$P^{-1}(k) = [\psi^{\mathrm{T}}(1)\,\psi^{\mathrm{T}}(2)\ldots\psi^{\mathrm{T}}(k)]\begin{bmatrix}\psi(1)\\ \psi(2)\\ \vdots\\ \psi(k)\end{bmatrix}$$

or

$$P(k) = \left[\sum_{i=1}^{k}\psi^{\mathrm{T}}(i)\psi(i)\right]^{-1}. \tag{7.36}$$

Similarly, substituting for Ψ_k and Y_k from equation (7.19) into equation (7.35) and simplifying yields the matrix $Q(k)$ as

$$Q(k) = \sum_{i=1}^{k}\psi^{\mathrm{T}}(i)y_i \tag{7.37}$$

where $y_i = y_i(nT)$. For m unknowns in the parameter vector $\hat{\theta}$ the matrices $P(k)$ and $Q(k)$ are of order $m \times m$ and $m \times 1$ respectively. It follows from equation (7.36) that

$$P^{-1}(k) = \sum_{i=1}^{k-1} \psi^{\mathrm{T}}(i)\psi(i) + \psi^{\mathrm{T}}(k)\psi(k)$$

or

$$P^{-1}(k) = P^{-1}(k-1) + \psi^{\mathrm{T}}(k)\psi(k).$$

Pre-multiplying this equation by $P(k)$, then post-multiplying by $P(k-1)$ yields

$$P(k-1) = P(k) + P(k)\psi^{\mathrm{T}}(k)\psi(k)P(k-1). \tag{7.38}$$

Post-multiplying equation (7.38) by $\psi^{\mathrm{T}}(k)$ and simplifying yields

$$P(k-1)\psi^{\mathrm{T}}(k) = P(k)\psi^{\mathrm{T}}(k)[1 + \psi(k)P(k-1)\psi^{\mathrm{T}}(k)]. \tag{7.39}$$

Post-multiplying equation (7.39) by $[1 + \psi(k)P(k-1)\psi^{\mathrm{T}}(k)]^{-1}\psi(k)P(k-1)$ yields

$$P(k-1)\psi^{\mathrm{T}}(k)[1 + \psi(k)P(k-1)\psi^{\mathrm{T}}(k)]^{-1} \times \psi(k)P(k-1) = P(k)\psi^{\mathrm{T}}(k)\psi(k)P(k-1).$$

Substituting for the right-hand side of the above equation from equation (7.38) and rearranging yields the covariance matrix at time step k as

$$P(k) = P(k-1) - P(k-1)\psi^{\mathrm{T}}(k)[1 + \psi(k)P(k-1)\psi^{\mathrm{T}}(k)]^{-1}\psi(k)P(k-1). \tag{7.40}$$

Similarly, using equation (7.37), the matrix $Q(k)$ can be written as

$$Q(k) = \sum_{i=1}^{k-1} \psi^{\mathrm{T}}(i)y_i + \psi^{\mathrm{T}}(k)y_k$$

or

$$Q(k) = Q(k-1) + \psi^{\mathrm{T}}(k)y_k. \tag{7.41}$$

Substituting for $P(k)$ and $Q(k)$ from equations (7.40) and (7.41) into equation (7.34) yields

$$\hat{\theta}(k) = \{P(k-1) - P(k-1)\psi^{\mathrm{T}}(k)[1 + \psi(k)P(k-1)\psi^{\mathrm{T}}(k)]^{-1}\psi(k)P(k-1)\} \times \{Q(k-1) +)\psi^{\mathrm{T}}(k)y(k)\}. \tag{7.42}$$

Simplifying equation (7.42) yields the parameter estimates at time step k as

$$\hat{\theta}(k) = \hat{\theta}(k-1) - P(k-1)\psi^{\mathrm{T}}(k)[1 + \psi(k)P(k-1)\psi^{\mathrm{T}}(k)]^{-1}\varepsilon(k) \tag{7.43}$$

where $P(k-1)Q(k-1) = \hat{\theta}(k-1)$ has been utilized and the error variable $\varepsilon(k)$ is defined as

$$\varepsilon(k) = \psi(k)\hat{\theta}(k-1) - y(k). \tag{7.44}$$

Substituting for $P(k-1)\psi^{T}(k)$ from equation (7.39) into equation (7.43) yields an alternative expression for the parameter estimates $\hat{\theta}(k)$ at time step k as

$$\hat{\theta}(k) = \hat{\theta}(k-1) - P(k)\psi^{T}(k)\varepsilon(k). \tag{7.45}$$

Equation (7.43), or equivalently equation (7.45), with equations (7.40) and (7.44) form the recursive least squares (RLS) estimator. Thus, the following two alternative forms of the RLS algorithm allow the updating of $\hat{\theta}(k)$.

Algorithm 7.1

1. Set the initial covariance matrix, $P(k-1)$, initial parameter vector $\hat{\theta}(k-1)$, and collect sufficient data to form the observation matrix $\psi(k-1)$.
2. Obtain input/output data to form the new observation matrix $\psi(k)$ and y_k.
3. Form the error variable $\varepsilon(k)$ using
$$\varepsilon(k) = \psi(k)\hat{\theta}(k-1) - y_k.$$
4. Form the covariance matrix $P(k)$ using
$$P(k) = P(k-1) - P(k-1)\psi^{T}(k)[1 + \psi(k)P(k-1)\psi^{T}(k)]^{-1}\psi(k)P(k-1).$$
5. Update the parameter estimates $\hat{\theta}(k)$ using
$$\hat{\theta}(k) = \hat{\theta}(k-1) - P(k)\psi^{T}(k)\varepsilon(k).$$
6. Set $k = k + 1$ and go to step 2.

Algorithm 7.2

1. Set the initial parameter estimates vector $\hat{\theta}(k-1)$ and initial covariance matrix $P(k-1)$ and form the observation matrix $\psi(k-1)$.
2. Obtain input/output data to form the new observation matrix $\psi(k)$ and y_k.
3. Form the error variable $\varepsilon(k)$ using
$$\varepsilon(k) = \psi(k)\hat{\theta}(k-1) - y_k.$$
4. Form the new parameter estimates $\hat{\theta}(k)$ using
$$\hat{\theta}(k) = \hat{\theta}(k-1) - P(k-1)\psi^{T}(k)[1 + \psi(k)P(k-1)\psi^{T}(k)]^{-1}\varepsilon(k).$$
5. Form the new covariance matrix $P(k)$ using
$$P(k) = P(k-1) - P(k-1)\psi^{T}(k)[1 + \psi(k)P(k-1)\psi^{T}(k)]^{-1}\psi(k)P(k-1).$$
6. Set $k = k + 1$ and go to step 2.

There is no apparent significant difference between Algorithm 7.1 and Algorithm 7.2. However, it follows from a comparison of the two algorithms that the former begins with less uncertainty about previous parameter estimates; by the time step (*v*) is executed elements of $P(k)$ would have already reduced in magnitude from their initial values. This means that a reduced level of effort will be put towards good adaptation. Therefore, the algorithm may not converge to the real parameter set θ. Algorithm 7.2 does not suffer from this problem and proves to be more robust.

A key variable in the RLS algorithm is the modelling error $\varepsilon(k)$ given in equation (7.44). This is the error between the predicted output using parameter estimates $\hat{\theta}(k-1)$ from the previous time step and the current system output $y_k(nT)$. For this reason it is usually called the prediction error or innovation. The true modelling error (or residual) at time step k would, however, be

$$e(k) = \psi(k)\hat{\theta}(k) - y_k. \tag{7.46}$$

The only difference between $\varepsilon(k)$ and $e(k)$ is that $\varepsilon(k)$ is based on the parameter estimates $\hat{\theta}(k-1)$ at time step $k-1$, whereas $e(k)$ is based on the parameter estimates $\hat{\theta}(k)$ at time step k. As the estimation time k increases the difference between $e(k)$ and $\varepsilon(k)$ becomes insignificant. However, during the early stages of the identification run the difference is significant. This difference is exploited by various estimation algorithms such as gradient methods.

7.3.4 Starting recursive least squares estimation

The RLS algorithm will provide the update or correction, at a time step k, to the previous parameter estimates $\hat{\theta}(k-1)$. Similarly, the calculation of $P(k)$ yields an update of the previous value $P(k-1)$. Therefore, to calculate the updates $\hat{\theta}(k)$ and $P(k)$ for time step k the parameter estimates vector $\hat{\theta}(k-1)$ and the covariance matrix $P(k-1)$ are needed from the previous time step. Initially, the values $\hat{\theta}(0)$ and $P(0)$ will not in general be known. Therefore, we require to give them reasonable values and then rely upon the linear quadratic correction mechanism of the RLS algorithm to perform subsequent estimation. Two methods of initializing the parameter estimates vector and covariance matrix are commonly used.

Prior linear equation fitting If there are m unknowns in the parameter estimates $\hat{\theta}$ then sufficient data can be collected for m time steps to form the data vectors Ψ_m and Y_m. Using equation (7.24), with the data vectors Ψ_m and Y_m, a set of m equations in m unknowns can be formed. The simultaneous solution of these equations yields $\hat{\theta}(m)$. Similarly, using equation (7.35) with the data vectors Ψ_m and Y_m will yield $P(m)$. The RLS estimation will thus start at time step $m+1$.

Guessing Alternatively, the initial parameter vector $\hat{\theta}(0)$ is formed by a reasonable guess to reflect the system dynamics, and the initial covariance matrix $P(0)$ is set so as to reflect the degree of uncertainty about the guess. A reasonable guess would be to set $\hat{\theta}(0)$ to correspond to a single discrete-time integrator, that is, $a_1(0) = 1$ and $b_1(0) = 1$ will all the other entries in $\hat{\theta}(0)$ equated to zero. An alternative, and more popular, guess is to set all the entries in $\theta(0)$ to zero.

$P(0)$ is usually chosen to be a diagonal matrix

$$P(0) = \rho I \tag{7.47}$$

where ρ is a real number. $P(0)$ corresponds to the degree of uncertainty in $\hat{\theta}(0)$ and sets the initial step length in the recursion. A large ρ corresponds to a high degree of uncertainty and thus will cause rapid fluctuations in the initial estimates. Conversely, a small ρ corresponds to a low degree of uncertainty and will lead to a slow adaptation of $\hat{\theta}(k)$. A $\rho = 10\,000$ is considered large whereas a value of $\rho = 100$ is considered small. The choice of ρ, besides reflecting a reasonable degree of uncertainty in the initial parameter vector, depends on a number of other factors, such as the word length of the digital computer and the accuracy of computation.

7.3.5 Time-varying parameters

The RLS algorithm minimizes the error criterion given in equation (7.21) and thus weights every error $e(i)$ equally. A change occurring in the system during an identification run requires that recent errors be more strongly emphasized than old errors. What is required is a means of allowing the RLS to adapt to the new system.

It follows from the previous discussion that RLS can be interpreted as a descent method, with the covariance matrix contributing to the step length information and direction of descent. After a sufficient number of iterations the parameter estimates converge and become more steady. This convergence is reflected as a decrease in the magnitude of the elements of the covariance matrix. Therefore, if the system parameters change, the ability of the recursive estimator to respond will be limited by the small magnitudes of the covariance matrix elements. One way to make the RLS respond to changes in the system is to increase the elements of $P(k)$ artificially at appropriate times. This is commonly done by either the method of random walks or the method of forgetting factors (Astrom and Hagglund 1984; Hagglund 1983; Salgado *et al.* 1988; Sanoff and Wellstead 1983; Tham and Mansoori 1988; Ydstie 1985).

Random walk The random walk approach increases the covariance matrix elements by adding a constant, positive-definite matrix R to it at a certain time step to allow the estimator to respond to a sensed change in the system.

Thus, when the identifier needs to correct the model of the system the covariance matrix is inflated:

$$P(k) = P(k) + R.$$

In this way individual elements of $P(k)$ can be separately affected by proper choice of R. Thus the random walk method is a more selective method of allowing for parameter variations. However, the application of this method requires more care.

Forgetting factor The forgetting factor approach is to build a finite memory into the RLS algorithm. Here, elements of the covariance matrix are scaled by a scalar parameter or 'forgetting factor' η (between 0 and 1) so that $P(k)$ is replaced by $P(k)\eta^{-1}$ throughout the RLS algorithm. When $\eta = 1$ (infinite memory) the normal algorithm applies. As η is decreased (finite memory), elements of the covariance matrix are slightly increased at each recursion allowing the algorithm to forget old parameter values, thus progressively reducing the weighting applied to past errors. The normal least squares criterion is to minimize, at each time step k, the cost function in equation (7.21):

$$J_n(k) = \sum_{i=1}^{k} e^2(i). \tag{7.48}$$

The forgetting factor approach is to minimize, at each time step k, a modified cost function J_f,

$$J_f(k) = \sum_{i=1}^{k} \eta^{k-i} e^2(i). \tag{7.49}$$

The influence of η is to reduce progressively the weighting given to old data. Commonly used values of η lie between 0.95 and 1.

The forgetting factor has the advantage that it is easy to implement in recursive form. It can be shown by working through the least squares derivation with the modified cost function of equation (7.49) that the modified least squares parameter estimates at time step k are

$$\hat{\theta}(k) = [\Psi_k^{\mathrm{T}} \eta(k) \Psi_k]^{-1} \Psi_k^{\mathrm{T}} \eta(k) Y_k \qquad (7.50$$

where $\eta(k)$ is the diagonal matrix

$$\eta(k) = \begin{bmatrix} \eta^k & 0 & . & . & . & 0 \\ 0 & \eta^{k-1} & . & . & . & 0 \\ \vdots & \vdots & \vdots & \vdots & \vdots & \vdots \\ 0 & 0 & . & . & . & 1 \end{bmatrix} \tag{7.51}$$

The introduction of a forgetting factor in recursive form is achieved by scaling-up the covariance matrix by $1/\eta$. The fact that it is implemented so easily in recursive form makes the forgetting factor technique very popular. However, it has the disadvantage of being indiscriminate in that it scales all elements of $P(k)$ by the same account. To localize the effect only the diagonal elements of $P(k)$ are sometimes scaled by $1/\eta$. This allows the use of different forgetting factors for each parameter, similar to the random walk method.

Under some time-varying conditions changes in system parameters may occur at unknown times and in between these times the parameters remain steady. In such cases an automatically varying forgetting factor is useful so that when the parameters are constant the forgetting factor is set to unity but it is allowed to drop when a change is detected. A lower forgetting factor will be desired on two occasions: at start-up and when a change is sensed during an identification run. At start-up, when the estimator has little or no knowledge of the system, it is useful to have a forgetting factor which starts small and increases towards unity at a prespecified rate. A variable forgetting factor which does this is given, at time step k, by

$$\eta_s(k) = \gamma + (1 - \gamma)[1 - e^{-k/\tau}] \tag{7.52}$$

where γ is the initial value of the forgetting factor (at $k = 0$) and τ is the time constant which determines the speed at which $\eta_s(k)$ will adapt. This form of forgetting factor is also used when the identifier is reset for whatever reason.

A change in the system, at time step k, during an identification run can be detected by a change in the modelling error $\varepsilon(k)$. If this error grows in magnitude it means that the identified model is incorrect and needs adjustment. At such times the forgetting factor is decreased to allow the model to adapt. A variable forgetting factor which allows this is given by

$$\eta_a(k) = \frac{\tau}{\tau - 1}\left[1 - \frac{\varepsilon^2(k)}{\tau v(k)}\right] \tag{7.53}$$

where τ determines the rate of adaptation of $\eta_a(k)$ and $v(k)$ is an average of the past values of $\varepsilon^2(k)$ given by

$$v(k) = \frac{\tau - 1}{\tau} v(k - 1) + \frac{\varepsilon^2(k)}{\tau}. \tag{7.54}$$

Substituting for $\varepsilon^2(k)$ from equation (7.54) into equation (7.53) and simplifying yields the adaptive forgetting factor as

$$\eta_a(k) = \frac{v(k - 1)}{v(k)}. \tag{7.55}$$

Combining the start-up/reset forgetting factor $\eta_s(k)$ of equation (7.52) with the adaptive forgetting factor $\eta_a(k)$ of equation (7.55) yields the variable forgetting factor $\eta_v(k)$. Thus

$$\eta_v(k) = \eta_s(k)\eta_a(k). \tag{7.56}$$

7.3.6 *Factors influencing the identification*

To identify an unknown process some knowledge of the process and signals is required. The a priori knowledge of the process is, for example, based on the general understanding of the process, on physical laws, and on prior measurements. Such a knowledge gives hints about whether the process is linear or non-linear, time-invariant or time-varying, the process's dominant time constants, time delays, and noise characteristics. The initial information provided to the identifier is based on this knowledge. The subsequent success of the identifier is influenced by a wide range of factors. A knowledge of these factors is important before starting an identification experiment. Therefore, some of these are briefly considered here.

Selection of input signals To identify a dynamic process the input signal has to satisfy certain conditions. A minimum requirement is that the input signal should persistently excite the dynamics of the process over the measurement period. This roughly means that the input signal has to be sufficiently rich to excite all process modes of interest during the experiment.

Selection of sampling time For the identification of discrete-time models the sampling time has to be properly selected before the experiment begins; if the signals are stored (as in self-tuning) the time constant cannot subsequently be changed. The sampling time depends primarily on:

(1) the final application; and
(2) the accuracy of the resulting model.

In self-tuning the resulting model from the identification process is used to design and implement a controller, therefore the selection of the sampling period is also influenced by the control algorithm, the spectrum of the signals and disturbances, and the computation burden on the digital processor.

The accuracy of the model is primarily influenced by the sampling period. With decreasing sampling periods the model exhibits higher-order behaviour and even small errors in the model parameters can have a significant influence on the input/output behaviour of the model. On the other hand, if the sampling period is too large the model looks like a constant or like multiple integrators and its dynamic behaviour will be too far from being exact. Therefore, the sampling time should neither be too small nor too large.

Model order The model order is one of the parameters which needs to be specified a priori to the identifier. It is known that if the model order is underestimated the identifier can drift into unacceptable regions leading to difficulties in convergence. If the model order $\hat{m}$ is specified smaller than the process order m than an approximate (sub-optimal) model is obtained. On the other hand, if $\hat{m}$ is greater than m then there will be $\hat{m} - m$ approximately cancelling (redundant) poles and zeros in the model transfer function. The existence of $\hat{m} - m$ redundant parameters in the updating process will obviously affect the processing speed and consume more memory space. Nevertheless, a model order $\hat{m} \geq m$ is preferred, depending on hardware and software limitations.

Initial model parameter vector The initial model parameter vector $\hat{\theta}(0)$ is set on the basis of a priori knowledge about the process. Care must be taken to choose $\hat{\theta}(0)$ properly. If in doubt, it is preferable to set all the entries in $\hat{\theta}(0)$ to zero.

Initial covariance matrix The initial covariance matrix elements are to reflect the degree of confidence in the initial parameter vector $\hat{\theta}(0)$. Normally the initial covariance matrix is set to ρI, where I is the identity matrix. A large value of ρ, indicating lack of confidence in the initial parameter vector $\hat{\theta}(0)$, leads to a rapid updating of $\hat{\theta}$, whereas a small value of ρ leads to slow changes in $\hat{\theta}$. The choice of ρ, normally chosen between 100 and 10 000, depends on the microprocessor's word length, its arithmetic computational power and on whether a slow or rapid convergence is desired.

Forgetting factor As the identification process is running the parameter estimates $\hat{\theta}$ converge to a steady set. This is reflected by a reduction in the magnitude of the elements of the covariance matrix P. If at this stage a change in the system occurs then a normal RLS algorithm will not be capable of correcting the model parameters $\hat{\theta}$ adequately to represent the process after the change. A commonly used approach to enable the updating of $\hat{\theta}$ after a change has occurred is to introduce a forgetting factor η into the RLS algorithm so that the covariance matrix is scaled up at each time step k; thus

$$P(k) = \Big[P(k-1) - P(k-1)\psi^{\mathrm{T}}(k) \times [1 + \psi(k)P(k-1)\psi^{\mathrm{T}}(k)]^{-1}\psi^{\mathrm{T}}(k)P(k-1)\Big]\frac{1}{\eta}. \quad (7.57)$$

Consider the case when the second term on the right-hand side of equation (7.57) is either zero or too small in magnitude, for example, the observation matrix is either zero or too small. Thus, equation (7.57) will reduce to

$$P(k) = P(k-1)\frac{1}{\eta}$$

and the covariance matrix will thus grow exponentially if $\eta < 1$. If there are no changes in $\psi(k)$ for a long period of time the matrix $P(k)$ may thus become too large. In this case a change in the command signal may then lead to large changes in the parameter estimates. The large value of $P(k)$ may also lead to numerical problems. This problem is known as the covariance windup and occurs whenever the vector $P(k-1)\psi^{T}(k)$ in equation (7.57) is either zero or sufficiently small over a period of time (Wittenmark and Astrom 1984). The problem is closely associated with identifiability conditions and the selection of the forgetting factor η.

An alternative to a constant forgetting factor is to use a variable forgetting factor η_v defined by equation (7.56). This will eliminate the covariance windup. Thirdly, selected elements of the covariance matrix can be changed (as in the random walk approach) when a change in the system is sensed. A fourth possibility is to stop the updating of $\hat{\theta}(k)$ and $P(k)$ when $P(k-1)\psi^{T}(k)$ is sufficiently small. A fifth possibility is to analyse the conditioning number of the covariance matrix P and to switch to a stochastic approximation algorithm when P becomes poorly conditioned.

It is worth mentioning that the stable estimation of parameters, besides the above, depends also on a number of other factors. Among these, the most important influential factors affecting the success of an identification exercise are the choice of output variables that can be measured, the precision of the measurements, the signal-to-noise ratio, the properties of the noise, the numerical procedure for calculating the estimates, redundancy of parameters in the process model, and computer word length.

7.3.7 *Real-time implementation*

The RLS estimation algorithm is implemented on a digital signal processor on the basis of known model order using the discrete-time model in equation (7.12).

A key feature required of a signal processor is the preservation of computational accuracy throughout the algorithm and a large dynamic range for representing data values, especially elements of the covariance matrix. In this respect a processor supporting floating-point arithmetic will be preferred. This aspect is important for the convergence of the parameter estimates vector $\hat{\theta}$ to the plant parameter vector θ.

With fixed-point arithmetic, where the computer word length is fixed and hence the dynamic range of computation is limited, the result of an arithmetic operation is accurate as long as no overflow occurs during the computational process. To avoid overflow as well as to increase the dynamic range, the implementation of floating-point arithmetic could be a favourable choice. However, this may not easily be achieved with a digital signal processor. An alternative is to split each data value into two (integer and decimal) parts and store each in a data location as an unscaled integer part and scaled decimal part. This can easily be achieved with a signal processor having instructions suitable for dealing with upper half and lower half bits of the accumulator either separately or together. Such a representation of numbers will thus require subroutines for computing the addition, subtraction, multiplication, and division of numbers. The extra computational effort required of the signal processor may lead, especially with high-order models, to limiting the sampling time and hence to imposing a limit on the maximum frequency of the input signal. However, if a low-order model is involved and/or the maximum frequency of the input signal is not high (as in ANC applications) this option will not lead to serious problems.

The initial covariance matrix $P(0) = \rho I$ is chosen to reflect the degree of confidence of the initial parameter estimates vector. A large ρ may normally be preferred to reflect a low degree of certainty about the parameter estimates vector and result in fast convergence. However, the value of ρ is limited by the processor dynamic range of computation. To avoid overflow problems ρ should be chosen such that the product $u_{\mathrm{m}}\rho$, where u_{m} is the maximum magnitude of the input sample, must be within the processor dynamic range of computation. It is possible to use a large value for ρ with a given u_{m} if the input signal is scaled down, provided the signal-to-noise ratio is kept within an acceptable level.

Note that the covariance matrix is symmetric about its diagonal elements. This means that it is sufficient to compute either the upper or lower half of the matrix only for updating the parameter estimates. In this manner a substantial amount of computational effort, such as time and storage memory, can be saved, especially with high-order models. Such savings are often important in practice where real-time signal processing is involved.

The frequency at which the input and output signals of a plant, to be identified, are sampled is normally determined by one full iteration cycle of the identification run. The sampling frequency in turn determines the upper limit of the signal frequency. In theory, the sampling frequency has to be at least twice the maximum frequency of the signal (Shannon's theorem). In practice, however, it is required that the sampling frequency be at least three times the maximum frequency of the input signal. Moreover, for stable convergence of the estimates $\hat{\theta}$ to θ, it is required that the sampling rates at both input and output ends of the plant are the same and uniform. Non-uniform or unequal sampling leads to unstable convergence and hence incorrect parameter estimates of the plant.

In an identification exercise, simultaneous sampling of the input and output signals of the unknown plant are required. Samples of the input and output signals are normally obtained through a process of A/D conversion. Thus, the samples are subject to delay during the A/D conversion process. Provided the delay at both input and output ends of the plant are equal, so that simultaneous sampling is achieved, a correct estimate of the plant model will be obtained. However, a difference between the two delays will introduce a phase lag in the estimated model with respect to the plant model. This is equivalent to the plant model, of transfer function $H(z)$, being in cascade with a delay element of constant amplitude and linear phase, so that the corresponding estimated model transfer function $\hat{H}(z)$ will be

$$\hat{H}(z) = z^{-\tau}H(z) \tag{7.58}$$

where τ is a non-negative real number. This delay element results if the A/D converters at the input and output ends of the plant are not identical, and in practice, even with identical A/Ds such a delay is possible.

7.4 Summary

An overview of the motivation and main techniques of adaptive control systems, in particular self-tuning control, and the recursive least squares identification algorithm has been given. Self-tuning control is an adaptive algorithm derived from the point of view of discrete-time stochastic control theory. It is characterized by the direct combination of an identification and a control algorithm so that the two processes proceed in a sequential manner.

To obtain a robust self-tuning control algorithm an accurate design procedure based on a good model is required. The key issues involved in obtaining a good model are valid data and an appropriate model structure based on a priori knowledge of the process. The models chosen are normally linear, in parametric form, and of low order. Therefore, the success of an identification exercise will depend on the properties of the input signal; the sampling period; the initial tuning parameters such as the covariance matrix and model parameters; and accuracy, computational power, and speed of the digital signal processor. To estimate an accurate reduced-order model it is essential that the input signal has sufficient energy content to fully excite all dynamic modes of interest of the plant, that is, the input signal should be persistently exciting.

The speed at which the input and output signals of a plant to be identified are sampled is as crucial as the properties of the input signal in the identification process. VLSI signal processors make implementation of digital signal processing (DSP) algorithms at high sampling rate feasible at reasonable cost. However, one of the main problems common to most DSP applications arise from the quantization of signals and coefficients, particularly in the case of fixed-point arithmetic.

The main feature desired of an identification algorithm is to be able to track variations in the process dynamics and provide a correct and up-to-date model. To do so it is necessary that the algorithm discounts old data. This will involve a compromise between discounting too quickly which will make the estimates uncertain if the parameters are constant and discounting too slowly which will make it impossible to track rapid parameter variations. The random walk and forgetting factors, accompanied with techniques of taking necessary precautions against consequent problems such as covariance windup, are the common methods of discarding old data and tracking variations in the process behaviour.

Provided the factors that crucially affect an identification process, such as model order, sampling period, initial covariance value, initial model parameter values, and type of input signal, are properly chosen the RLS identifier will result in an estimated process model that will have input/output characteristics very close to the unknown process. The estimated model may not necessarily have the same parameter values as the unknown plant even if the model order is the same as the process order. However, it will exhibit the same input/output behaviour as the unknown plant.

7.5 References

Astrom, K.J. (1970). *Introduction to stochastic control theory*. Academic Press, New York.

Astrom, K.J. (1983*a*). Theory and applications of adaptive control–A survey. *Automatica*, **19**, (5), 471–86.

Astrom, K.J. (1983*b*). LQG self-tuners. *Proceedings of IFAC Workshop on Adaptive Systems in Control and Signal Processing*, San Francisco, pp. 137–46.

Astrom, K.J. and Eykhoff, P. (1971). System identification–A survey. *Automatica*, **7**, (2), 123–62.

Astrom, K.J. and Hagglund, T. (1983). Automatic tuning of simple regulators for phase and amplitude margins specifications. *Proceedings of IFAC Workshop on Adaptive Systems in Control and Signal Processing*, San Francisco, pp. 271–6.

Astrom, K.J. and Hagglund, T. (1984). Automatic tuning of simple regulators with specifications on phase and amplitude margins. *Automatica*, **20**, (5), 645–51.

Astrom, K.J. and Wittenmark, B. (1973). On self-tuning regulators. *Automatica*, **9**, (2), 185–99.

Astrom, K.J. and Wittenmark, B. (1980). Self-tuning controllers based on pole-zero placement. *IEE Proceedings*, Part D, **127**, (3), 120–30.

Astrom, K.J. and Wittenmark, B. (1985). The self-tuning regulators revisited. *Preprints of 7th IFAC/IFORS Symposium on Identification and System Parameter Estimation*, York, **1**, xxv–xxxiii. Pergamon, Oxford.

Astrom, K.J. and Wittenmark, B. (1989). *Adaptive control*. Addison-Wesley, Reading.

Astrom, K.J. and Zhao-Ying, Z. (1982). Linear quadratic Gaussian self-tuner. *Proceedings of IEE Workshop on Adaptive Control*, Florence, pp. 1–20.

Astrom, K.J., Borisson, U., Ljung, L., and Wittenmark, B. (1977). Theory and applications of self-tuning regulators. *Automatica*, **13**, (5), 457–76.

Bohm, J. (1985). LQ self-tuners with signal level constraints. *Preprints of 7th IFAC/IFORS Symposium on Identification and System Parameter Estimation*, York, **1**, 799–802. Pergamon, Oxford.

Brown, R.G. (1983). *Introduction to random signal analysis and Kalman filtering*. John Wiley, New York.

Clarke, D.W. and H-James, R. (1971). Design of digital controllers for randomly disturbed systems. *Proceedings of IEE*, **118**, (10), 1503–6.

CLARKE, D.W. and GAWTHROP, P.J. (1975). A self-tuning controller. *Proceedings of IEE*, **122**, (9), 929–34.

CLARKE, D.W. and GAWTHROP, P.J. (1979). Self-tuning control. *Proceedings of IEE*, **126**, (6), 633–40.

EGARDT, B. (1980). Unification of some discrete-time adaptive control schemes. *IEEE Transactions on Automatic Control*, **AC-25**, (4), 693–7.

EYKHOFF, P. (1981). *Trends and progress in system identification.* Pergamon Press, Oxford.

GAWTHROP, P.J. (1977). Some interpretations of the self-tuning controller. *Proceedings of the IEE (Control & Science)*, **124**, (10), 889–94.

GAWTHROP, P.J. (1980). Hybrid self-tuning control. *Proceedings of IEE*, Part D, **127**, (5), 229–36.

GOODWIN, G.C., RAMADGE, P.J., and CAINES, P.E. (1980). Discrete-multivariable adaptive control. *IEEE Transactions on Automatic Control*, **AC-25**, (3), 449–56.

HAGGLUND, T. (1983). The problem of forgetting old data in recursive estimation. *Proceedings of IFAC Workshop on Adaptive Systems in Control and Signal Processing*, San Francisco, pp. 213–14.

HARRIS, C.J. and BILLINGS, S.A. (1981). *Self-tuning and adaptive control: theory and applications.* Peter Peregrinus, London.

IONNOU, P. (1983). Robust redesign of adaptive control in the presence of disturbances and unmodelled dynamics. *Proceedings of IFAC Workshop on Adaptive Systems in Control and Signal Processing*, San Francisco, pp. 19–25.

ISERMANN, R. (1982). Parameter adaptive control algorithms–A tutorial. *Automatica*, **18**, (5), 513–28.

ISERMANN, R. (1985). Parameter adaptive control systems–A review of methods and applications. *Preprints of 7th IFAC/IFORS Symposium on Identification and System Parameter Estimation*, York, **1**, 373–8. Pergamon, Oxford.

KALMAN, R.E. (1958). Design of a self-optimizing control system. *Transactions of the American Society of Mechanical Engineers*, **80**, 468–78.

KOSUT, R.L. (1983). Robustness issues in adaptive control. *Proceedings of IFAC Workshop on Adaptive Systems in Control and Signal Processing*, San Francisco, pp. 13–18.

KURZ, H., ISERMANN, R., and SCHUMANN, R. (1980). Experimental comparison and application of various parameter adaptive control algorithms. *Automatica*, **16**, (2), 117–33.

LANDAU, Y.D. (1979). *Adaptive control–The model reference approach.* Marcel Dekker, New York.

LJUNG, L. and SODERSTROM, T. (1983). *Theory and practice of recursive identification.* The Massachusetts Institute of Technology Press, Cambridge, Mass.

MISHKIN, E. and BRAUN, L. JR. (1961). *Adaptive control systems.* McGraw-Hill, New York.

NAGY, I. and PETERKA, V. (1985). A hybrid LQ self-tuning controller. *Preprints of 7th IFAC/IFORS Symposium on Identification and System Parameter Estimation*, York, **1**, 1025–31. Pergamon, Oxford.

NARENDRA, K.S. and LIN, Y. (1980). Stable discrete-adaptive control. *IEEE Transactions on Automatic Control*, **AC-25**, (3), 456–61.

NARENDRA, K.S., LIN, Y., and VALAVANI, L.S. (1980). Stable adaptive controller design. *IEEE Transactions on Automatic Control*, **AC-25**, (3), 440–8.

PETERKA, V. (1970). Adaptive digital regulation of noisy systems. *IFAC 2nd Symposium on Identification and Process Parameter Estimation*, Prague, pp. 6.2/1–11.

PRALY, L. (1983). Robustness of indirect adaptive control based on pole placement design. *Proceedings of IFAC Workshop in Adaptive Systems in Control and Signal Processing*, San Francisco, pp. 55–60.

QIN, Z. (1985). A generalized minimum variance self-tuning controller with pole-assignment. *Preprints of 7th IFAC/IFORS Symposium on Identification and System Parameter Estimation*, York, **1**, 403–8. Pergamon, Oxford.

ROHRS, C.E., VALAVANI, L., ATHANS, M., and STEIN, G. (1982). Robustness of adaptive control algorithms in the presence of unmodelled dynamics. *Proceedings of 21st IEEE Conference on Decision and Control*, Florida, pp. 3–11.

SALGADO, M.E., GOODWIN, G.C., and MIDDLETON, R.H. (1988). Modified least squares algorithm incorporating exponential resetting and forgetting. *International Journal of Control*, **47**, (2), 477–91.

SANOFF, P. and WELLSTEAD, P.E. (1983). Implementation of self-tuning regulators with variable forgetting factors. *Automatica*, **19**, (3), 345–6.

STREJC, V. (1979). Trends in identification. *Proceedings of 5th IFAC Symposium on Identification and System Parameter Estimation*, Darmstadt, **1**, 1–16.

STREJC, V. (1980). Least squares parameter estimation. *Automatica*, **16**, 535–50.

THAM, M.T. and MANSOORI, S.N. (1988). Covariance resetting in recursive least squares estimation. *Proceedings of IEE International Conference on Control*, Oxford, pp. 128–33.

WELLSTEAD, P.E. and SANOFF, S.P. (1981). Extended self-tuning algorithm. *International Journal of Control*, **34**, (3), 433–55.

WELLSTEAD, P.E. and ZANKER, P.M. (1979). Servo self-tuners. *International Journal of Control*, **30**, (1), 27–36.

WELLSTEAD, P.E., PRAGER, D., and ZANKER, P.M. (1979a). Pole-assignment self-tuning regulator. *Proceedings of IEE*, **126**, (8), 781–7.

WELLSTEAD, P.E., EDMUNDS, J.M., PRAGER, D., and ZANKER, P.M. (1979b).Self-tuning pole/zero assignment regulators. *International Journal of Control*, **30**, (1), 1–26.

WITTENMARK, B. (1974). Self-tuning predictor. *IEEE Transactions on Automatic Control*, **AC-19**, (6), 848–51.

WITTENMARK, B. (1975). Stochastic adaptive control methods – A survey. *International Journal of Control*, **21**, (5), 705–30.

WITTENMARK, B. and ASTROM, K.J. (1984). Practical issues in the implementation of self-tuning control. *Automatica*, **20**, (5), 595–605.

XUE-SHAN, Y. (1985). Implicit extended pole-assignment self-tuning controller. *Preprints of 7th IFAC/IFORS Symposium on Identification and System Parameter Estimation*, York, **2**, 1787–91. Pergamon, Oxford.

YAMANE, Y., NIKIFORUK, P.N., and GUPTA, M.M. (1983). Deadbeat adaptive control in feedback. *Proceedings of IFAC Workshop on Adaptive Systems in Control and Signal Processing*, San Francisco, pp. 97–8.

YDSTIE, B.E. (1985). Adaptive control and estimation with forgetting factors. *Preprints of 7th IFAC/IFORS Symposium on Identification and System Parameter Estimation*, York, **2**, 1761–6. Pergamon, Oxford.

8

SELF-TUNING ACTIVE NOISE CONTROL

8.1 Introduction

An ANC system employing a controller with a fixed frequency-dependent transfer function performs well in situations where the characteristic behaviour of the system does not vary. However, if the system is subject to variation, for example due to changes in the operating conditions of the source or changes in geometry, then satisfactory performance is not achieved with an ANC system employing a controller with fixed transfer characteristics. Indeed, it may result in a reinforcement in the overall level of noise. In such circumstances an adaptive controller is required so that its characteristics are updated in accordance with changes in the system so that the desired level of noise cancellation is achieved and maintained.

It follows from the theory of adaptive control presented in Chapter 7 that the design of an adaptive controller for use in ANC systems involves two processes;

(1) modelling; where a suitable model for the system is obtained using a system identification algorithm such as the RLS algorithm;
(2) controller design calculation; where the controller parameters are computed on the basis of the identified model and system performance objectives.

In this manner, it is first required to identify signals as system input and output between which a suitable model can be obtained. Moreover, the identification of such a model should lead to a suitable controller design formulation. In an ANC system the primary signal at the source and the observed signal will ideally constitute the system input and output, respectively.

Measurement of the observed signal can easily be achieved by using an acoustic sensor (observer microphone). On the other hand, measurement of the primary signal at the source is not easy in practice, because no direct access to this signal will normally be available. However, the required information on this signal can be obtained through a detection process:

(1) a direct measurement of the acoustic signal at a distance from the source; for example using a detector microphone;

(2) an indirect measurement of the signal by a vibration sensor, for example.

In this manner, the detected signal can be used as the system input, thus attempting to obtain a model describing the system between the detection and observation points.

In direct detection, the detector will in general give a combined measure of the primary and secondary signals reaching the detection point. This leads to a partial measure of the primary source signal. The use of a directional sensor can eliminate, to some extent, the effect of secondary source radiation reaching the detector. However, due to a requirement of a minimum separation between the sources, total elimination of the effect of secondary source radiation reaching the detection point is difficult to achieve in practice. Therefore, such an effect must be taken into account in the process of modelling the system.

In indirect detection, as discussed earlier, necessary information on the acoustic signal is obtained through a measurement of a non-acoustic coherent signal. Here, the measurement is not in general affected by secondary source radiation. However, care is required to ensure that full information about the primary noise spectrum is obtained.

A self-tuning ANC algorithm can be realized by a proper integration of a system identification algorithm with the controller design rules. The flexibility of the self-tuning design strategy makes it possible to develop such an algorithm within an ANC system's framework, thereby utilizing the self-tuning property to achieve desired cancellation of the unwanted noise. Moreover, such an algorithm can be implemented effectively and economically using VLSI signal processing devices.

This chapter presents and discusses the design methodology and implementation of a self-tuning ANC algorithm for three-dimensional propagation. Frequency domain mathematical models describing the system between the detection and observation points are obtained. The controller transfer function required for optimum cancellation of noise is related to characteristics of the system models, leading to the desired controller design rules. Processes of system identification and controller design are integrated so as to result in a self-tuning control algorithm suitable for on-line implementation. Practical realization of the algorithm is discussed and its significance verified through a noise cancellation exercise of a practical source.

8.2 The process model

Consider the ANC system in the general form of an FFCS (see Fig. 4.3). The corresponding block diagram of this, in the complex frequency s domain, is shown in Fig. 4.4. This block diagram is shown in a slightly modified form in

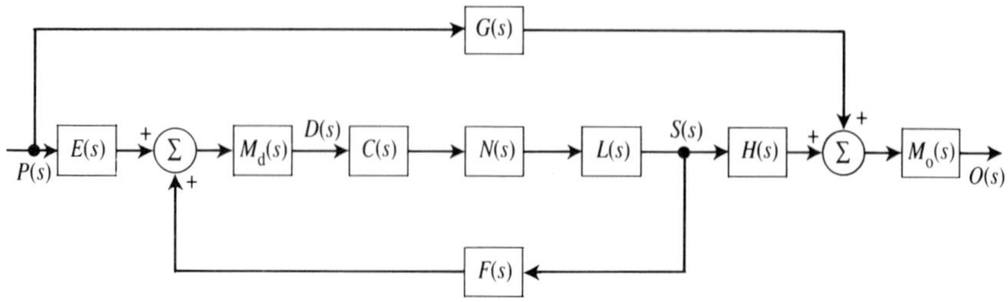

Fig. 8.1. Block diagram of ANC system

Fig. 8.1, where:

$D(s)$ = signal measured by the detector,
$O(s)$ = signal measured by the observer,
$M_d(s)$ = transfer function of the detector,
$M_o(s)$ = transfer function of the observer,

and all the rest of the quantities are the same as in Fig. 4.4.

An ANC system can, in general, be considered as having two signal generating or source points and two measurement points, with the primary and secondary sources constituting the source points and the detector and observer constituting the measurement points. Thus the detected and observed signals $D(s)$ and $O(s)$ can each be thought of as composed of the sum of two signals, one due to the primary source and one due to the secondary source. From the block diagram in Fig. 8.1 this can be expressed as

$$\begin{aligned} D(s) &= M_d(s)E(s)P(s) + M_d(s)F(s)S(s) \\ O(s) &= M_o(s)G(s)P(s) + M_o(s)H(s)S(s). \end{aligned} \tag{8.1}$$

The secondary source signal $S(s)$, on the other hand, can be expressed, via the controller path, as

$$S(s) = C(s)N(s)L(s)D(s). \tag{8.2}$$

Note that, in an ANC system the primary source operates independently and entirely outside the control bounds of the system. The secondary source, on the other hand, is derived from the primary source and hence its operation is under full control of the system. This feature leads to identifying two distinct situations; namely, when the secondary source is off and when the secondary source is on.

8.2.1 Secondary source off

When the secondary source is off the signal measured at the detection and observation points are due to the primary source only. This situation is

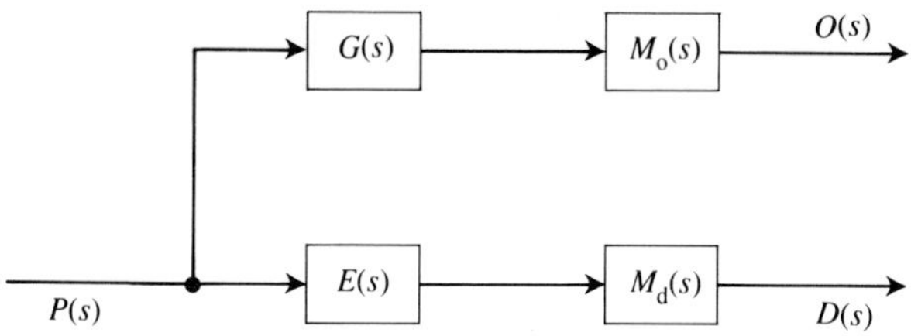

Fig. 8.2. System block diagram with secondary source off

characterized by the secondary source signal $S(s)$ in Fig. 8.1 being zero:

$$S(s) = 0. \tag{8.3}$$

This can be achieved by disconnecting the link between the detector and the secondary source via the controller path. Thus, Fig. 8.1 reduces to the equivalent block diagram shown in Fig. 8.2. This block diagram, or equivalently, substituting for $S(s)$ from equation (8.3) into equation (8.1) yields the detector and observer signals $D(s)$ and $O(s)$ as

$$\begin{aligned} D(s) &= M_d(s)E(s)P(s) \\ O(s) &= M_o(s)G(s)P(s). \end{aligned} \tag{8.4}$$

Dividing equations (8.4) side by side yields the equivalent transfer function describing the system between the detection and observation points as

$$\frac{O(s)}{D(s)} = \frac{M_o(s)G(s)}{M_d(s)E(s)}; \qquad S(s) = 0.$$

Let this transfer function be denoted by $X_0(s)$; thus

$$X_0(s) = \frac{O(s)}{D(s)} \qquad \text{for } S(s) = 0 \tag{8.5}$$

or

$$X_0(s) = \frac{M_o(s)G(s)}{M_d(s)E(s)}. \tag{8.6}$$

It follows from the above that if the secondary source is off then the subsystem with $D(s)$ as input and $O(s)$ as output can be described by the simplified block diagram equivalently shown in Fig. 8.3.

8.2.2 Secondary source on

When the secondary source is on then the signals measured at the detection and observation points will be the result of a combined effect of the primary

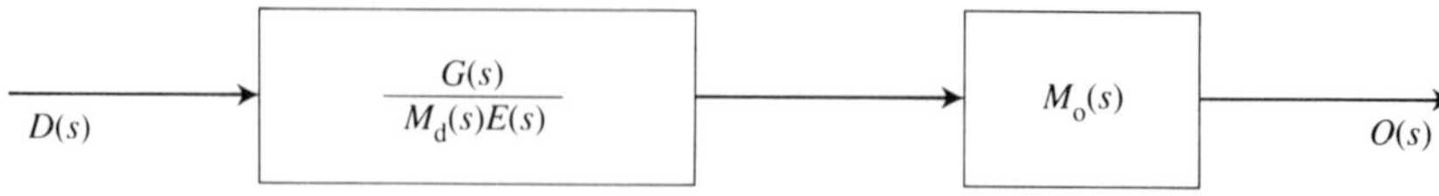

Fig. 8.3. Equivalent system block diagram with secondary source off

and secondary signals at these points. This situation is characterized by the secondary source signal $S(s)$ in Fig. 8.1 not being zero:

$$S(s) \neq 0. \tag{8.7}$$

This can be achieved by providing a link between the detector and secondary source via the controller path. Since the controller transfer function is not known at this stage it is reasonable to assume that the link is provided by closing a switch. This is equivalent to replacing $C(s)$ in Fig. 8.1 by a block with unity gain and zero phase. Alternatively, a general connecting block of a particular transfer function may be assumed. In either case $C(s)$ in Fig. 8.1 can be replaced by unity and the characteristic effect of the link can be accounted for within the transfer function $N(s)$. Thus, Fig. 8.1 leads to the equivalent block diagram shown in Fig. 8.4.

The block diagram in Fig. 8.4 or, equivalently, substituting $C(s) = 1$ into equation (8.2) yields

$$S(s) = N(s)L(s)D(s). \tag{8.8}$$

Substituting for $S(s)$ from equation (8.8) into equations (8.1) yields

$$\begin{aligned} D(s) &= M_d(s)E(s)P(s) + M_d(s)N(s)L(s)F(s)D(s) \\ O(s) &= M_o(s)G(s)P(s) + M_o(s)N(s)L(s)H(s)D(s). \end{aligned} \tag{8.9}$$

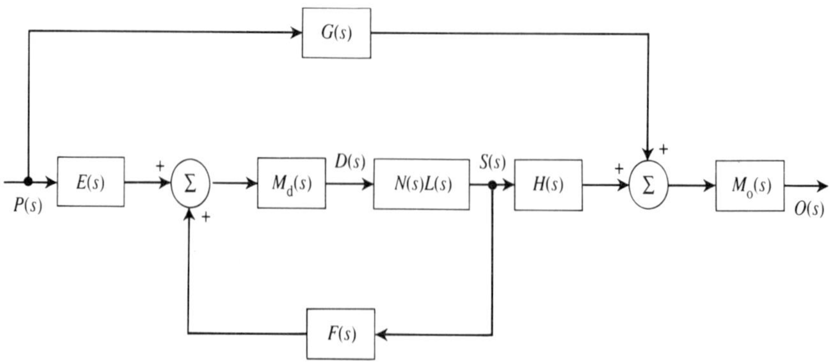

Fig. 8.4. System block diagram with secondary source on

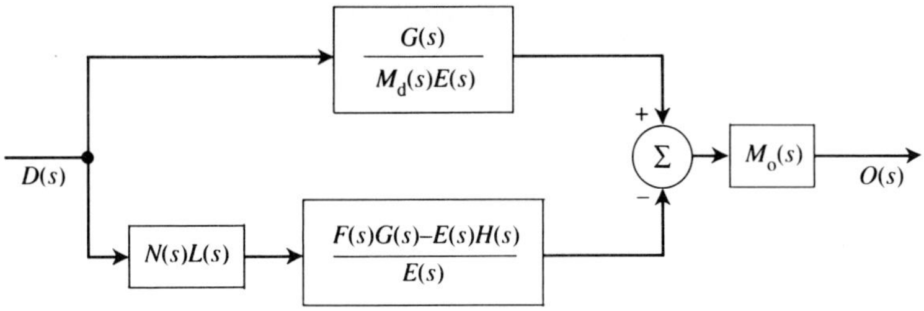

Fig. 8.5. Equivalent system block diagram with secondary source on

Eliminating $P(s)$ in equations (8.9) and simplifying yields the equivalent transfer function describing the system between the detection and observation points as

$$\frac{O(s)}{D(s)}=\frac{M_o(s)G(s)}{M_d(s)E(s)}-M_o(s)N(s)L(s)\left[\frac{G(s)F(s)-E(s)H(s)}{E(s)}\right]; \qquad S(s)\neq 0.$$

Let this transfer function be denoted by $X_1(s)$; thus

$$X_1(s)=\frac{O(s)}{D(s)} \qquad \text{for } S(s)\neq 0 \tag{8.10}$$

or

$$X_1(s)=\frac{M_o(s)G(s)}{M_d(s)E(s)}-M_o(s)N(s)L(s)\left[\frac{G(s)F(s)-E(s)H(s)}{E(s)}\right]. \tag{8.11}$$

Therefore, if the secondary source is on, then the subsystem with $D(s)$ as input and $O(s)$ as output can equivalently be described by the simplified block diagram shown in Fig. 8.5.

8.3 Controller design

Models describing the ANC system were obtained in the previous section. These are given by the transfer functions $X_0(s)$ in equation (8.6) and $X_1(s)$ in equation (8.10). In this section design rules for the calculation of the required controller transfer function $C(s)$ are obtained in terms of the transfer functions $X_0(s)$ and $X_1(s)$.

Simplifying equation (8.11) yields

$$X_1(s)=\frac{M_o(s)G(s)}{M_d(s)E(s)}\left[1-M_d(s)N(s)L(s)\left(\frac{G(s)F(s)-E(s)H(s)}{G(s)}\right)\right].$$

Note that the ratio $M_o(s)G(s)/[M_d(s)E(s)]$ is the transfer function $X_0(s)$, given by equation (8.6), and the second term within the square brackets in the above equation is the inverse of the required controller transfer function $C(s)$; see equation (4.13). Thus

$$X_1(s) = X_0(s)\left[1 - \frac{1}{C(s)}\right]. \tag{8.12}$$

Solving equation (8.12) for $C(s)$ yields the required controller design rule in terms of the transfer characteristics of the system models as

$$C(s) = \left[1 - \frac{X_1(s)}{X_0(s)}\right]^{-1} \tag{8.13}$$

The controller design rule thus obtained is based on a continuous-time description of the system. For discrete implementation of the design rule, corresponding relations in the discrete-time domain are required. This can simply be achieved by applying an appropriate s- to z-domain transformation method to the above relations. Thus, the corresponding discrete-time equivalent transfer functions are

$$C(z) = \left[1 - \frac{X_1(z)}{X_0(z)}\right]^{-1} \tag{8.14}$$

where

$$X_0(z) = \frac{M_o(z)G(z)}{M_d(z)E(z)}$$

$$X_1(z) = \frac{M_o(z)G(z)}{M_d(z)E(z)} - M_o(z)N(z)L(z)\left[\frac{G(z)F(z) - E(z)H(z)}{G(z)}\right]. \tag{8.15}$$

Equation (8.14) is the required discrete controller design rule given in terms of the discrete-time system model transfer functions $X_0(z)$ and $X_1(z)$.

8.4 Controller identification

The controller design rule developed in the previous section is based on the assumption that the transfer functions of the system models are known. In practice, these are not initially known. To obtain the required controller transfer function, system models described by $X_0(z)$ and $X_1(z)$ have to be identified first. This is done by using an identification algorithm, such as the RLS estimation.

The RLS algorithm gives estimates of parameters of the system model on the basis of a known model structure and measurements of the system input and output. The model describing the system can be assumed to be of the form

$$O(nT) = \sum_{i=0}^{m} b_i D(nT - iT) + \sum_{i=1}^{m} a_i O(nT - iT) + g(nT) \qquad (8.16)$$

where $D(nT)$ as system input and $O(nT)$ as system output are, respectively, the current samples of the detected and observed signals, $g(nT)$ is an uncorrelated white noise, m is the order of the model, T is the sampling time, and n is an integer real number. The identifier will give estimates of the parameters a_i and b_i of the process model for known model order m and suitable sampling period T under each situation described by Figs. 8.3 and 8.5. In this manner, the identification process will proceed in two phases, once to estimate $X_0(z)$ and then to estimate $X_1(z)$, thus resulting in the corresponding estimated transfer functions $\hat{X}_0(z)$ and $\hat{X}_1(z)$. The switching of identification from one phase to the other can be performed automatically under program control. Such an automatic switching is desirable for on-line controller design.

An important factor to be considered carefully in the two phases of identification is the sampling period. Since the estimated transfer functions $\hat{X}_0(z)$ and $\hat{X}_1(z)$ are subsequently used in the controller design calculation, care must be taken to ensure that the sampling times in both identification phases are consistent. If the sampling times are different then the resulting controller transfer function will be incorrect for either sampling time.

The model order is chosen on the basis of practical knowledge about the system behaviour. If the model order is overestimated, that is, $\hat{m} > m$, there will be unnecessary redundant parameters in the model which will limit the computational speed and available memory space for data storage within the digital signal processor. On the other hand, if the model order is underestimated, $\hat{m} < m$, the identifier will converge to an incorrect model. Thus, the consequences of underestimation of the model order are potentially more severe than overestimation.

8.4.1 Controller design calculation

Under the condition where the secondary source is off, the ANC system is equivalent to that shown in Fig. 8.3 with the corresponding discrete-time transfer function given by $X_0(z)$ in equation (8.15). This transfer function can be thought of as the ratio $M_o(z)/M_d(z)$ of transfer functions of the observer and detector cascaded with the ratio $G(z)/E(z)$ of transfer functions of the acoustic paths through distances r_g and r_e. If the detector and observer (microphones) are identical then $X_0(z)$ will be dominantly determined by the ratio $G(z)/E(z)$.

Provided the primary source satisfies the identifiability conditions, the RLS algorithm can be used to estimate parameters of $X_0(z)$ using the process model in equation (8.16) with a suitable $m = m_0$. If $m_0 = 1$, for instance, then the transfer function $X_0(z)$ can be expressed as

$$X_0(z) = \frac{b_{01} + b_{11}z^{-1}}{1 - a_{11}z^{-1}} \tag{8.17}$$

where $\{a_{11}, b_{i1}\}$, for $i = 0, 1$, are the process model parameters.

Under the condition where the secondary source is on, the ANC system is equivalent to that shown in Fig. 8.5 with the corresponding discrete-time transfer function given by $X_1(z)$ in equations (8.15).

Using the RLS algorithm the parameters of the transfer function $X_1(z)$ can be estimated, assuming a model of the form in equation (8.16) with a suitable $m = m_1$, provided the identifiability conditions due to the input signal are satisfied. Assuming a second-order process model, for instance, the transfer function $X_1(z)$ can be expressed as

$$X_1(z) = \frac{b_{02} + b_{12}z^{-1} + b_{22}z^{-2}}{1 - a_{12}z^{-1} - a_{22}z^{-2}} \tag{8.18}$$

where $\{a_{i2}, b_{j2}\}$, for $i = 1, 2$ and $j = 0, 1, 2$, are the process model parameters for this case.

Once the process transfer functions $X_0(z)$ and $X_1(z)$ have been identified, equation (8.14) can be used to calculate the corresponding controller transfer function. Thus, substituting for $X_0(z)$ and $X_1(z)$ from equations (8.17) and (8.18) into equation (8.14) yields

$$C(z) = \left[1 - \frac{(b_{02} + b_{12}z^{-1} + b_{22}z^{-2})(1 - a_{11}z^{-1})}{(1 - a_{12}z^{-1} - a_{22}z^{-2})(b_{01} + b_{11}z^{-1})}\right]^{-1}.$$

Simplifying the above yields the controller transfer function as

$$C(z) = \frac{b_{0c} + b_{1c}z^{-1} + b_{2c}z^{-2} + b_{3c}z^{-3}}{1 - a_{1c}z^{-1} - a_{2c}z^{-2} - a_{3c}z^{-3}} \tag{8.19}$$

where the controller parameters are

$$\begin{aligned}
b_{0c} &= \frac{b_{01}}{b_{01} - b_{02}} & & \\
b_{1c} &= \frac{b_{11} - b_{01}a_{12}}{b_{01} - b_{02}} & a_{1c} &= -\frac{b_{11} - b_{12} - b_{01}a_{12} + b_{02}a_{11}}{b_{01} - b_{02}} \\
b_{2c} &= -\frac{b_{01}a_{22} + b_{11}a_{12}}{b_{01} - b_{02}} & a_{2c} &= \frac{b_{01}a_{22} + b_{11}a_{12} + b_{22} - b_{12}a_{11}}{a_{01} - a_{02}} \\
b_{3c} &= -\frac{b_{11}a_{22}}{b_{01} - b_{02}} & a_{3c} &= \frac{b_{11}a_{22} - a_{11}b_{22}}{b_{01} - b_{02}}.
\end{aligned} \tag{8.20}$$

Equations (8.20) are the controller design rules given on the basis of a first-order process model describing the system when the secondary source is off and a second-order process model describing the system when the secondary source is on. The model orders were chosen arbitrarily in the controller design procedure described above. Depending on the application the model orders can be different from the above. However, the design procedure will be the same.

8.4.2 *Controller design algorithm*

The controller design procedure outlined in the previous section can be formulated algorithmically as follows.

Algorithm 8.1

1. Switch off the secondary source.
2. Use the RLS algorithm to obtain the estimated model transfer function $\hat{X}_0(z)$.
3. Switch on the secondary source.
4. Use the RLS algorithm to obtain the estimated model transfer function $\hat{X}_1(z)$.
5. Use $\hat{X}_0(z)$ and $\hat{X}_1(z)$ obtained in steps 2 and 4 to calculate an estimate of the required controller transfer function $\hat{C}(z)$ using

$$\hat{C}(z) = \left[1 - \frac{\hat{X}_1(z)}{\hat{X}_0(z)}\right]^{-1}. \tag{8.21}$$

The above algorithm is described schematically in Fig. 8.6 where the block labelled 'plant' represents the system as seen between the detector and observer under either steps 1 and 2 or steps 3 and 4.

Note that since the identification of $X_0(z)$ and $X_1(z)$ can be performed independently of one another, either of these can be estimated first. This, in terms of Algorithm 8.1, means that an equivalent algorithm is obtained by interchanging steps 3 and 4 with steps 1 and 2, that is, first estimating $X_1(z)$ and then $X_0(z)$.

To achieve automatic switching of the secondary source a subroutine within the identification program will be required so that the execution of the algorithm proceeds without manual intervention. This can easily be done using an input/output channel of the digital signal processor (on which Algorithm 8.1 is implemented) as a connecting block in the controller path, under program control, so that when $S(z) = 0$ is required the output is kept at zero, whereas when $S(z) \neq 0$ is required, data is transferred to the output as

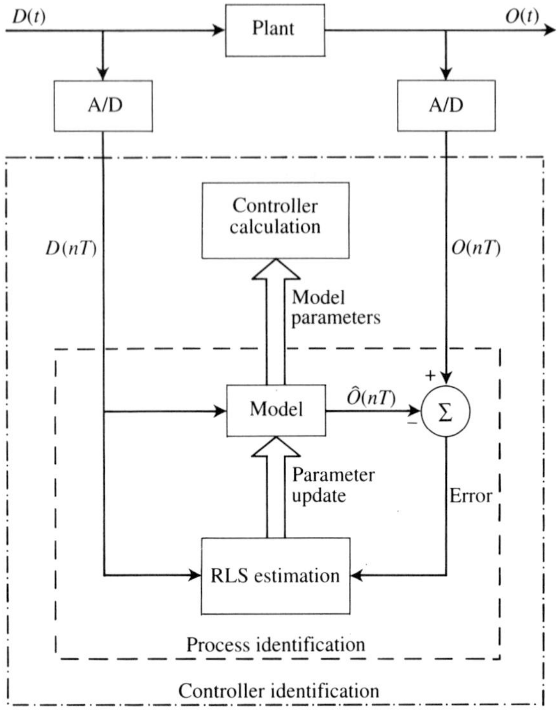

Fig. 8.6. Process of controller identification

received. This will effectively implement an automatic switch in the controller path. The identification can then proceed as follows.

1. Switch off the secondary source, that is, do not output the data as received.
2. Activate the RLS program to estimate $X_0(z)$.
3. Switch on the secondary source, that is, output the data as received.
4. Activate the RLS algorithm to estimate $X_1(z)$.

Once the estimates $\hat{X}_0(z)$ and $\hat{X}_1(z)$ have been obtained in this manner the controller design calculation can proceed accordingly. This process is described schematically in Fig. 8.7 where the switch SW is in position *a* during the estimation of $X_0(z)$ and moves to position *b* when estimation of $X_1(z)$ commences.

Note that for the process of automatic switching from one phase of identification to the next a criterion is required on the basis of which the identifier is commanded when to switch to the next phase. One method is to specify a number indicating the total number of iterations the RLS algorithm would require for a reasonable convergence of the parameter estimates to the

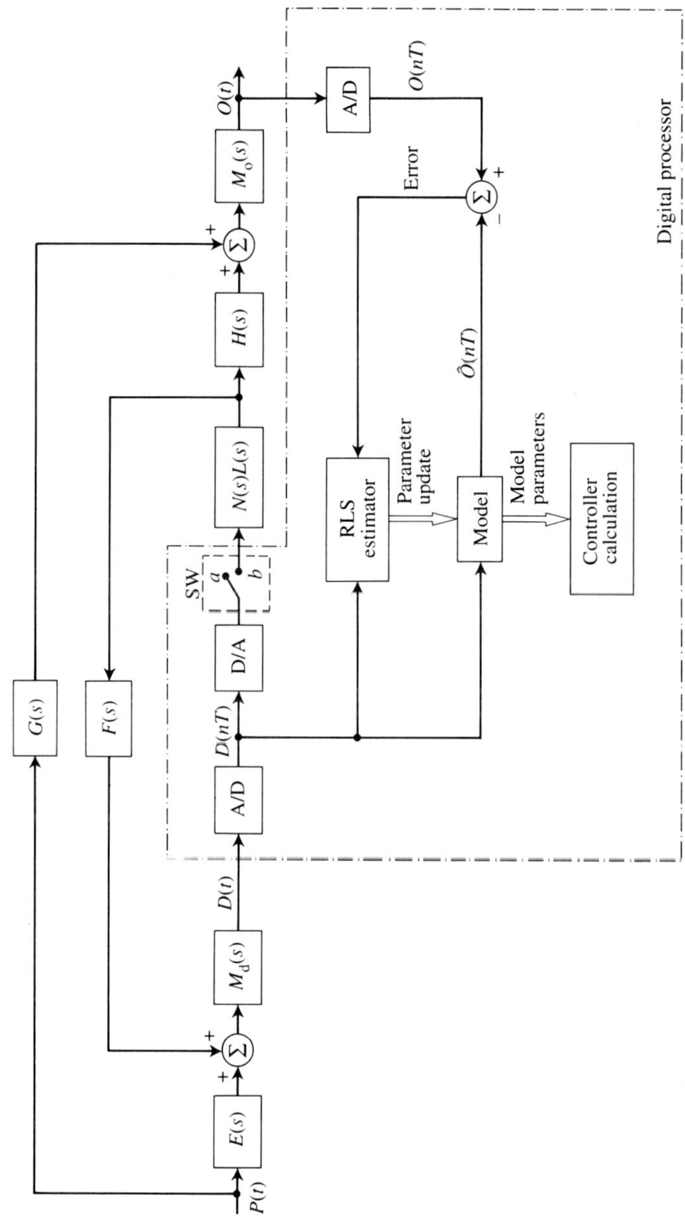

Fig. 8.7. Process of controller design

plant parameters. Once the specified number of iterations has been completed the identifier can switch to the next phase. This method requires some practical experience with the RLS algorithm for the application. A second method is to use the magnitude of the covariance matrix elements. Here the diagonal elements of the covariance matrix can give a measure of convergence, so that when these are reasonably small in magnitude the identifier will proceed to the next phase. A third possibility is to use the magnitude of the modelling error $\varepsilon(k)$ as a measure of convergence, so that when $\varepsilon(k)$ is reasonably small in magnitude the identifier will proceed to the next phase. A quantitative measure of 'reasonably small' in the second and third methods depends on how close a convergence is desired. Both the second and third methods give a measure of performance of the identifier and their use does not require as much practical knowledge as the first method. Moreover, either of the second and third methods can also provide a means of controlling the algorithm during the course of an identification run. For instance, if the covariance matrix elements or the modelling error decrease in magnitude and eventually a reasonably good convergence is achieved the algorithm can proceed to the next phase. However, if the covariance matrix elements or the modelling error grow in magnitude the algorithm is diverging and hence the identification run should restart.

The model orders for $X_0(z)$ and $X_1(z)$ are chosen on the basis of a knowledge about the system and the processing speed limitations of the digital signal processor. A choice of $\hat{m}_0$ for $X_0(z)$ and $\hat{m}_1$ for $X_1(z)$ will lead to a controller transfer function of order $\hat{m}_0 + \hat{m}_1$. The model order, as discussed in Chapter 7, has a direct effect on the sampling time. Therefore, care must be taken to ensure that the sampling period remains consistent throughout the identification process. With a consistent sampling time the identifier will give good estimates of both $X_0(z)$ and $X_1(z)$ and hence give a controller that will exhibit the required response characteristics. However, if the sampling time is not consistent throughout the identification process an incorrect controller will result.

8.4.3 Controller implementation

The adaptive controller algorithm as derived and formulated above can be implemented on a digital signal processor to result in on-line design and implementation of the controller as a digital controller. Some of the practical issues of importance at this stage are briefly discussed in this subsection.

It was noted and discussed in Chapter 7 that due to a difference in the delays in A/D converters at the input and output ends of the plant a constant delay is introduced into the estimated model transfer function. As far as the implementation of Algorithm 8.1 is concerned, this problem does not appear crucial. If such a delay is introduced it will appear, with the same value, in both $\hat{X}_0(z)$ and $\hat{X}_1(z)$ and hence will effectively cancel during the calculation of the ratio $\hat{X}_1(z)/\hat{X}_0(z)$ in equation (8.21). Thus, the controller transfer function will not be affected by this error.

In implementing the controller transfer function, a linear delay in the phase of the controller is introduced due to the A/D and D/A conversion circuitry of the digital signal processor. This is equivalent to the implemented transfer function cascaded with the transfer function of the A/D and D/A circuitry. To obtain the correct phase characteristics a compensating network is commonly implemented in cascade with the transfer function. The compensating network is designed to have a constant amplitude characteristic and a rising phase as a function of frequency within the frequency range of the input signal. The required characteristic behaviour can thus be achieved by either implementing the compensating network externally using an analogue network of a suitable transfer function or internally within the processor using a discrete-time transfer function.

8.5 Design of the self-tuning control algorithm

A self-tuning ANC combines the complementary processes of identification and control in such a manner that the two will proceed sequentially. Moreover, a supervisory level of adaptation is used to monitor system performance and activate self-tuning control.

The process of identification employs the RLS algorithm and consists of estimating models of the system under the two situations considered in Section 8.2; namely, when the secondary source is off and when it is on. This results in estimates of the transfer functions $X_0(z)$ and $X_1(z)$. The process of control, on the other hand, consists of the controller design calculation and the implementation of the controller transfer function, as considered in Section 8.4.

It follows from the above that the process of control can only commence after the transfer functions $X_0(z)$ and $X_1(z)$ have been fully estimated. In many applications of self-tuning control, a controlled process with a single input/output behaviour is involved. Here, an initial controller is used at the start of self-tuning, in which case, as the process proceeds the controller is adapted to the required transfer function. In such cases the controller design calculation is thus dependent on a (single) model of the controlled process. Moreover, in these applications the self-tuning adaptation mechanism relies on parameters within the identification process and no performance-related measure is given outside the identification and control processes to monitor system performance and activate self-tuning control accordingly. An ANC system, as noticed in Section 8.2, gives rise to two distinct kinds of subsystem behaviour due to the state of the secondary source. This leads to two distinct system models. Therefore, a self-tuning control algorithm within an ANC system's framework will require the full estimation of system models before the commencement of the control process. Moreover, to properly integrate the processes of identification and control as well as to monitor system performance and activate self-tuning accordingly a meta-level of control is required. This can be realized by designing a supervisory level of adaptation on the basis of a performance criterion so as to continually monitor system performance against the criterion and take proper action accordingly. Thus, the self-tuning ANC as a combined identification and control algorithm with a supervisory level control can be described schematically as in Fig. 8.8. This can be realized by incorporating a supervisor within Algorithm 8.1 resulting in the following algorithm.

Algorithm 8.2

1. Set the controller transfer function to zero; $C(z) = 0$.
2. Use the RLS identifier to obtain the estimated system model transfer function $\hat{X}_0(z)$.
3. Set the controller transfer function to unity; $C(z) = 1$.
4. Use the RLS identifier to obtain the estimated system model transfer function $\hat{X}_1(z)$.

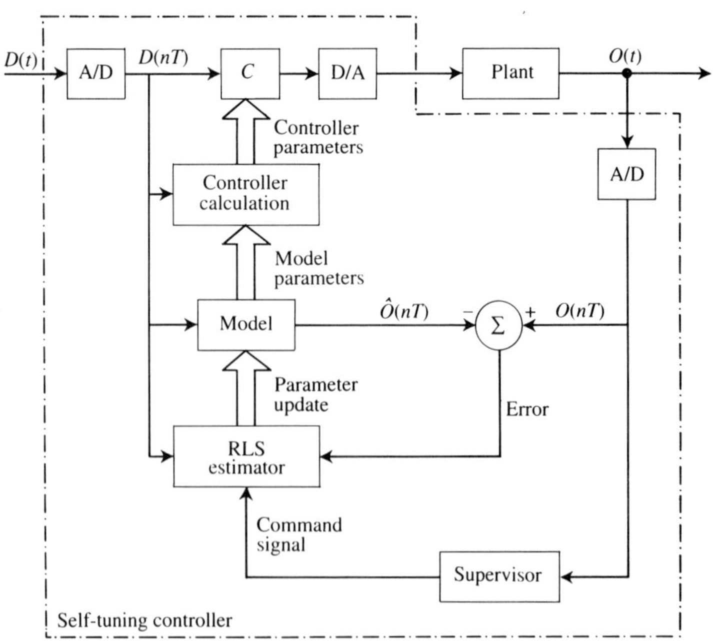

Fig. 8.8 Self-tuning controller

5. Use the estimated transfer functions $\hat{X}_0(z)$ and $\hat{X}_1(z)$ to calculate an estimate of the required controller transfer function $\hat{C}(z)$ using

$$\hat{C}(z) = \frac{1}{1 - \dfrac{\hat{X}_1(z)}{\hat{X}_0(z)}}.$$

6. Set and realize $\hat{C}(z)$ as a digital controller.
7. Measure the system performance and compare with the specified index.
8. If the system performance is within the specified range then go to step 7 otherwise go to step 1.

The index of performance used as a criterion by the supervisor will normally be an indicator of the level of noise cancellation achieved either in terms of reduction in average power or average pressure level. However, the absolute level of noise, for example, can also be used. In either case the level of performance desired can be specified quantitatively by the user. The supervisor thus obtains a quantitative measure of the actual level of system performance at each sample time through a measurement of the observed signal and compares this with the specified index.

The supervisor can also act as a monitor of the performance of the identifier, as a controller stability monitor, or as a system stability monitor. As a monitor of the performance of the identifier such factors as redundant poles and zeros due to an overestimation of model order, and the nature of the estimated transfer function, for example, minimum or non-minimum phase, can be identified. As a controller stability monitor the supervisor can identify unstable poles, if any, in the estimated controller transfer function. Furthermore, as discussed in Chapter 5, owing to system geometry and the acoustic feedback loop the system may become unstable. Such a situation can be identified and avoided through a measurement of either the detector signal or the controller output signal. Therefore, the more intelligence that is incorporated within the supervisor the more robust the algorithm will be.

Implementation of the self-tuning active noise control algorithm on a digital signal processor will result in the self-tuning ANC system shown in Fig. 8.9, where the block labelled 'controller' represents Algorithm 8.2 or, equivalently, the broken-line block in Fig. 8.8.

8.6 Implementation of the self-tuning control algorithm

The self-tuning ANC algorithm derived in the previous section has been presented in a manner that is suitable and easy to implement on a digital signal processor. This, however, requires a careful consideration of the

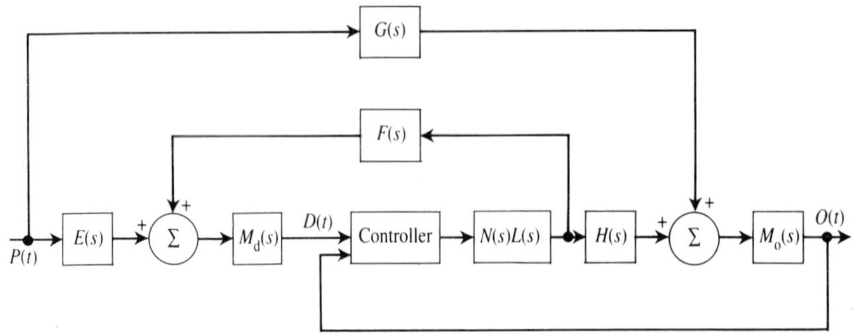

Fig. 8.9. Self-tuning active noise control system

practical issues that affect the implementation process and hence the performance of the controller. These, which are also common to the implementation of almost all digital controllers, include factors such as properties of the input signal, robustness of the estimation process, initialization of parameters at the start, and sampling time; and processor-related factors such as signal quantization, computational power, speed of computation, programming support, and data format. These are discussed briefly below in relation to the implementation of the self-tuning ANC algorithm.

8.6.1 Input signal conditioning

In all digital control applications it is important to have a proper conditioning of the signals. Due to the aliasing problem connected with the sampling process it is necessary to remove all frequencies above the Nyquist frequency before sampling the signals. High frequencies may otherwise be interpreted as low frequencies and may introduce aliases into the system. The filtering of the signals also has the effect that the system will be excited by frequencies where it is important to have good process models.

In digital signal processors the aliasing problem is commonly eliminated by a low-pass (anti-aliasing) filter that band-limits the input signal before sampling. The anti-aliasing filter usually has a flexible cut-off frequency which can be adjusted so that the incoming analogue signal is band-limited to the required maximum frequency before sampling. In ANC applications a disadvantage of this filter, however, is the introduction of an additional phase lag in the path of the controller that should be compensated for within the controller transfer function. With a high cut-off frequency the effect of this additional phase lag is not significant. However, with a low cut-off frequency this additional phase will lead to a high-order controller transfer function.

8.6.2 Robustness of the algorithm

A robust control algorithm requires an accurate design procedure based on a good process model. To obtain a good model, on the other hand, requires good data and an appropriate model structure. The models used are invariably simple, for example, linear and of low order. Therefore, in such a case, the estimates obtained will depend crucially on the properties of the input signal. To estimate an accurate reduced-order model it is essential that the input signal has sufficient energy and is persistently exciting over the frequency range of interest. The conditions on persistent excitation are related to the complexity of the estimated model. This implies that the requirements on the input signal become more severe if the model order is increased.

In ANC applications the excitation is provided by the unwanted (primary) source of noise. In simulation experiments where the primary source can be chosen properly to suit the robustness requirements of the control algorithm the problems due to the properties of the input signal will not arise. In practice, however, care must be taken to condition the input signal properly before sampling. The process model order is a second important factor influencing the accuracy of the model and hence the robustness of the control algorithm. Underestimation of the model order leads to inaccurate estimates and hence unsatisfactory performance. Overestimation of the model order, on the other hand, leads to redundant, approximately cancelling, poles and zeros in the estimated transfer function. This can have two consequences: (a) the redundant parameters will take extra computational time and hence increase the sampling time and (b) at high sampling frequencies the difference between the redundant poles and zeros, which are numerically close to each other, could be significant and result in an incorrect estimated model.

The key property of an adaptive controller is the ability to track variations in the process dynamics and to do so it is necessary to discount old data. The speed at which data is discounted needs to be chosen properly. If the parameters are constant and the data is discounted too fast the estimates will become uncertain. On the other hand, if the discounting of data is too slow, this will make it impossible to track rapid parameter variations. The method of the exponential forgetting factor is one way to discard old data and works well only if the process is properly excited at all times. There are several problems with exponential forgetting when the excitation of the process changes. A typical situation is when the main source of excitation changes in the set point (d.c. level). Then there may be long periods with no excitation and the estimator will discount old information and the uncertainties of the parameters will grow (estimation windup). Estimation windup can be avoided by using one of the several ways discussed in Chapter 7.

8.6.3 Tuning parameters

There are basically two categories of parameters to be chosen in a self-tuning ANC application: process-related and performance-related. The process-related parameters are typically the order of the process model and possibly the time-delay in the controlled process. The performance-related parameters are the performance indices and initial values for the estimation routine. The process-related parameters are usually easy to determine. The performance-related parameters, however, require some understanding of the system behaviour and can be chosen such that the performance indices are insensitive to the initial values of the estimator.

One parameter that can greatly influence the behaviour of the algorithm is the sampling period. The choice of the sampling interval is not specific to adaptive controllers but is an important design parameter for all sampled data design methods. The controller in a self-tuning ANC system requires estimates of the transfer functions $X_0(z)$ and $X_1(z)$ to be accurate and consistent with the same sampling frequency. This can be achieved by properly choosing the orders of $X_0(z)$ and $X_1(z)$ and making sure the sampling frequency remains constant throughout the identification and control processes. The model orders have a significant effect on the sampling time. An increased order taking more computational time and memory space will reduce the sampling frequency significantly. This will consequently limit the maximum frequency of the input signal and can even make the controller impractical to use.

8.6.4 Data quantization

A general problem with discretized controllers is that the zero-order hold (ZOH) at the outputs introduces considerable phase lag. Thus discretized controller frequency responses are likely either to show a more negative phase compared to the continuous controller, or to show increased gain in the higher frequency region, which stems from the attempt to lift the phase by introducing a compensating transfer function in cascade with the controller. Stability and damping problems may also occur. In ANC applications, where the accuracy requirements of the phase characteristics of the digital controller are crucial, the phase delay introduced by the ZOH can have a significant effect on the performance of the system. This requires the use of a compensating transfer function in cascade with the controller. The introduction of a delay compensation will lead to an increase in high-frequency controller gain and can cause instability in the system. However, this problem can be avoided by a proper choice of a compensating transfer function and/or band-limiting the controller output.

8.6.5 Processor word length and data format

Because of the limited number range due to finite word lengths, care must be taken that data, for instance controller states and coefficients, will fit into the processor word length. Numbers should not exceed the range, but at the same time should not be so small that the quantization effects become dominant. The product of two numbers is of almost double length and if the accumulator is double length, which is often the case, the products are accumulated in full length and a truncating or rounding operation is performed on the final result. This introduces a quantization error into the computations. Addition and subtraction of numbers, on the other hand, are exact as long as no overflow occurs.

The arithmetic data format can be as crucial as speed. In many cases floating-point arithmetic is desirable, be it because the dynamic range required is indeed large, or because the implementation does not want to deal with the problems of fixed-point arithmetic. Fixed-point numbers have a much more limited dynamic range than floating-point numbers with comparable word length. Moreover, with fixed-point numbers problems of overflow can easily occur.

When using fixed-point arithmetic it is usually necessary, at least, to perform scaling on the controller to be implemented. The primary objective is to fit data which are computed during the course of a difference equation calculation into the limited number range, so that overflows are avoided without provoking excessive signal quantization effects. A second objective with scaling is to alter coefficients in such a way that they fit into the coefficient number range. This is not always achieved when scaling is only oriented towards data overflow avoidance and scale factors then have to be altered appropriately.

The accuracy requirements of the estimated models and consequently of the controller in implementing the self-tuning ANC algorithm would require a large dynamic range. One way to achieve this with a processor supporting fixed-point arithmetic is to implement floating-point arithmetic within the processor. An alternative method, as discussed in Chapter 7, is to separate numbers into two parts: scaled integer and unscaled decimal. Both these methods have consequent effects of increased computational time, limiting memory space, and so on. The effect of increased computational time which affects the sampling time can be crucial if high-order process models are involved.

8.6.6 Controller realization structure

After estimating the processor models and calculating the controller transfer function the controller is implemented as a digital filter on the processor. In this phase of the implementation, to avoid problems of overflow and minimize quantization effects, a suitable single-input single-output (SISO) structure for the controller realization is required. Various realization structures of practical importance are known in digital filtering. Among these, the most commonly used SISO structures are the direct, cascade, and parallel structures (Antoniou 1979; Oppenheim *et al.* 1984; Terrel 1980).

In a direct structure realization of a transfer function the polynomials of the transfer function appear directly as coefficients in the difference equation or block diagram. It is well known that direct structures suffer from various drawbacks. First, the coefficients can easily be spread over a large number range, causing problems with number representation and arithmetic. This is because sums, products, and sums of products of polynomial roots form the coefficients and roots can be anywhere from the origin even to outside the unit circle in the z-plane. Second, the sensitivity of the roots or coefficient errors can be large. Such errors are introduced due to coefficient quantization in the processor. Another related drawback is potentially high gain sensitivity. Assuming a stable transfer function, a direct structure, involving the quantized versions of the coefficients, is likely to introduce inaccurate small differences of large numbers because the coefficients are frequently of large absolute value with alternating signs. Finally, direct structures suffer from particularly high signal quantization noise, which relates to high coefficient sensitivity. From the above it follows that direct structures should only be recommended for low-order systems or subsections of higher-order systems and should be used with care.

A reliable SISO structure, which is well accepted in digital filtering, is the cascade structure, where the transfer function is implemented in factorized form as a series connection of low-order blocks, usually of first- or second-order. This structure offers the possibility of optimal distribution of poles and zeros among the blocks, and internal block structures can be chosen optimally. However, its drawback for control applications is that the structure introduces increased computational delay. This is because the output appears only after computation in every block is completed.

A structure which is widely regarded to be as good a candidate as the cascade structure is the parallel structure. It corresponds to implementing the transfer function in a partial fraction expansion form. The partial fraction blocks are commonly chosen to be of first- and second-order and can be implemented by suitable structures. An advantage of parallel structures over the cascade structures is that they can be used in multiple-input multiple-output (MIMO) cases.

A careful consideration of the main issues discussed above in the process of implementing the self-tuning ANC algorithm will ensure a robust controller that can be used successfully for the cancellation of unwanted broadband noise.

8.7 Investigation of system performance

In this section an example of the performance of the self-tuning controller in the cancellation of noise of a practical source is presented. The source of noise is the exhaust noise of the motorcycle considered in Chapter 6. The self-tuning controller is implemented in a digital signal processor supporting fixed-point arithmetic. A compensating transfer function is implemented in cascade with the controller transfer function to compensate for the phase shift introduced due to the processor A/D and D/A circuitry. The supervisory level of adaptation activates self-tuning on the basis of a prespecified amount of noise cancellation. The input signal is band-limited to a maximum frequency of 200 Hz determined by the sampling time set by the processor. System components are typically arranged so that the observer is nearly 1.5 metres away and equidistant from the primary and secondary sources.

With the motorcycle engine running at an idle speed the cancellation as a function of frequency will typically be as shown in Fig. 8.10. Note that the 0 dB line partitions the regions of cancellation (below the line) and reinforcement (above the line). Here, the noise is cancelled over a broad frequency range of about 8 Hz to the 200 Hz maximum frequency of the input. The average amount of cancellation achieved is more than 10 dB (a prespecified value). The reinforcement at frequencies below 8 Hz is due to large attenuation in the secondary source (loudspeaker).

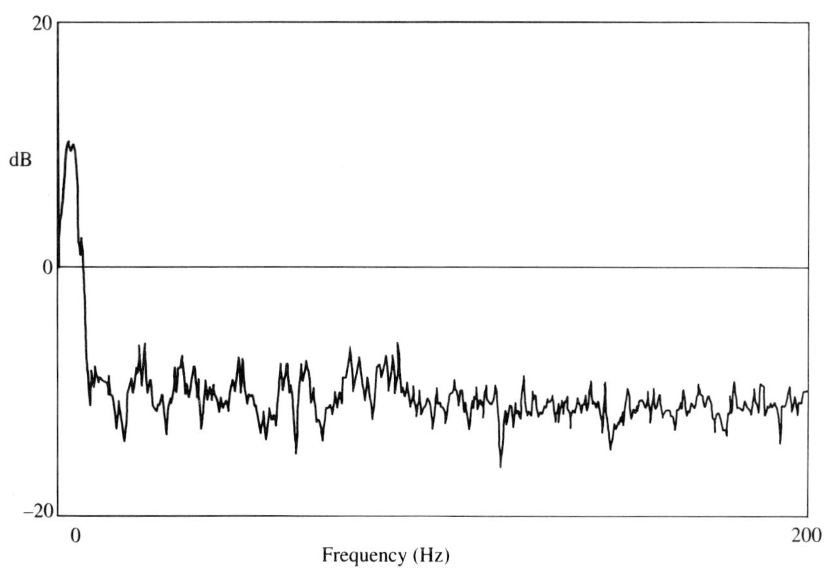

Fig. 8.10. Cancellation as function of frequency with the motorcycle engine running at idle speed

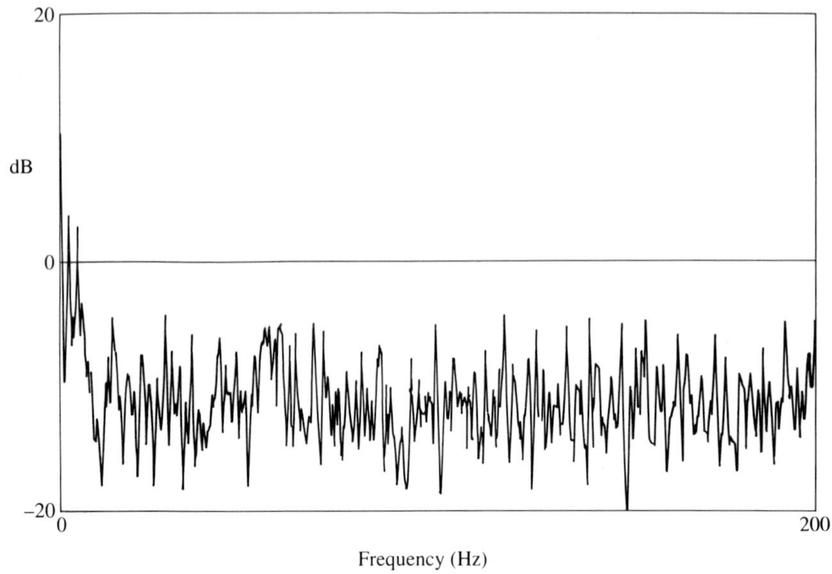

Fig. 8.11. Cancellation as function of frequency with the motorcycle engine running at a medium speed

With the motorcycle engine running at a medium speed the cancellation as a function of frequency will typically be as shown in Fig. 8.11. Here, the noise is cancelled over a broad frequency range of almost the lowest noise frequency to the 200 Hz maximum frequency of the input. The average amount of cancellation achieved is more than 10 dB. The sharp variations noticed in Fig. 8.11 are due to the fast varying nature of the noise spectrum in this case. Such variations are not dominantly evident in Fig. 8.10 where the engine is running at an idle speed.

The cancellation as a function of frequency shown in either Fig. 8.10 or Fig. 8.11 is characterized by a steady and large amount over a broad frequency range. As compared with the fixed controller in Chapter 6, the self-tuning ANC system, having the ability to update the controller characteristics in accordance with changes in the noise spectrum and geometry of the system, results in cancellation of the noise from almost the minimum to maximum noise frequency. This example verifies the practical significance of an adaptive ANC system for cancellation of broadband noise under time-varying conditions. Thus, the self-tuning ANC algorithm implemented on a dedicated digital signal processor will be able to perform well and to within practically acceptable requirements. A change in the system leading to a degradation of system performance is detected by the supervisor, upon which the self-tuning control is reset so that the desired level of cancellation is achieved and maintained.

8.8 Summary

A procedure for the design of an adaptive controller for use in ANC systems in three-dimensional propagation has been derived and its implementation discussed. The design and implementation of a self-tuning active noise control algorithm has been presented, discussed, and its performance investigated. The active cancellation of broadband noise, to an optimal level, requires a controller with continuous frequency-dependent characteristics that satisfy the system optimal performance criterion.

An ANC system employing a controller with fixed transfer characteristics will perform to a satisfactory level provided variations in the system do not occur. However, if the system is subject to variation, for example, due to operating conditions of the source or changes in transducer characteristics or system geometry, the system will not perform to a desired level. In these circumstances an adaptive ANC system is required so that the controller characteristics are updated in accordance with changes in the system and the desired level of performance is achieved and maintained.

The design of an adaptive controller requires a suitable model of the controlled process in question and a performance criterion. System identification algorithms, such as the RLS, give estimates of the system model in parametric form on the basis of a known model structure and measurements of system input and output signals.

An ANC system can be thought of as composed of two source points and two measurement points; the detection and observation points comprise the measurement points. Thus the detector and observer signals are used, respectively, as system input and output, and models of the system are estimated accordingly.

The primary and secondary sources comprise the source points in an ANC system. Among these, the primary source operates independently and outside the control bounds of the ANC system. The secondary source, on the other hand, is derived from the primary source through a process of detection and control and is thus under full control of the ANC system. This feature leads to two distinct situations; namely, when the secondary source is off and when the secondary source is on. Thus models describing the system behaviour under the two situations are obtained. The dependence of the required controller characteristics on the transfer characteristics of the two subsystems makes it possible to formulate a controller design algorithm that meets the time-varying requirements of an ANC system. A proper integration of the controller design with the system identification algorithm thus leads to a self-tuning ANC algorithm that can easily be implemented on a digital signal processor and used successfully for the cancellation of broadband noise under time-varying conditions.

8.9 References

Antoniou, A. (1979). *Digital filters: analysis and design.* McGraw-Hill, New York.

Oppenheim, A.V., Willsky, A.S., and Young, I.T. (1984). *Signals and systems.* Prentice-Hall, Englewood Cliffs.

Terrel, T.J. (1980). *Introduction to digital filters.* Macmillan, London.

9
PROSPECTS FOR ACTIVE NOISE CONTROL

The possibility of active noise control has intrigued scientists and engineers in many countries for over five decades. The early efforts, prior to the mid-1960s, were unsuccessful. In retrospect, this was due to a restricted understanding of the problem from an acoustic and control systems point of view, as well as the limitations in technology to be able to deal with the main problems associated with the design of ANC systems.

Advances in technology which began in the mid-1960s, and are continuing, make the active control of noise much more feasible and attractive. These include advanced control systems technology, which includes the development of adaptive control algorithms; improved understanding of the physics of acoustic systems; and the development of dedicated digital signal processing devices, making sophisticated control schemes realizable and inexpensive.

Previous developments of ANC systems have tended to focus on overcoming the practical problems of system implementation by developing complex geometrical structures: multiple-detector and/or multiple-loudspeaker configurations. However, only limited success has been attained. Recent advances in semiconductor technology have produced extremely fast, dedicated, digital signal processing devices capable of implementing algorithms of significant complexity in real-time. The opportunity now exists to trade geometric complexity with controller complexity, resulting in more robust designs.

An analysis of ANC systems from a control engineering perspective involves an understanding of the characteristic nature of the sources concerned and accounting for the acoustic response of the propagation medium and electronic equipment used in the system. Three basic characteristics of noise sources are defined:

(1) broadband sources;
(2) distributed sources;
(3) time-varying sources.

Consideration of each of the above leads to a specific control solution, and the most general case, comprising a combination of the three, will require complex control analysis. The limited success of previous attempts at developing ANC systems can be seen as a failure to account for one or more of the above characteristics. Furthermore,

(4) acoustic response of the medium; and
(5) response characteristics of transducers and other electronic equipment

each play an important role in the performance of an ANC system, and require careful consideration during the design stage. These, together with the geometric arrangement of the system, can result in poor performance and possibly sound reinforcement or even instability in the system. Moreover, inside an enclosure, the multi-mode response of the enclosure has an important effect on the performance of an ANC system.

The design of a robust ANC system requires an investigation of the problem from an acoustics and control engineering perspective. The implementation of such an approach leads to a clear description and understanding of the process of noise cancellation by active means and consequently to a suitable design methodology. The process of active noise cancellation relies on the interference of acoustic waves so that when a secondary source of noise is introduced destructive interference occurs leading to a reduction of the unwanted sound. This is achieved by artificially generating a secondary source of noise that is identical to the primary source and produces a wave that is in anti-phase with the unwanted noise so that when the two waves are superimposed on each other the noise is attenuated. In this process, the introduction of a secondary source results in a redistribution of the acoustic energy in the medium so that noise is reduced in some areas while reinforced in others. Thus, a pattern of zones of cancellation and reinforcement is created in the medium. Moreover, if the secondary source is synchronized with the unwanted noise then some of the acoustic energy will be absorbed by the secondary source, so the secondary source will be acting as an acoustic absorber.

The extent of zones of cancellation is dependent on the separation between the sources which, in turn, determines the required power ratio and the frequency of the component waves. Full cancellation of broadband noise over the entire three-dimensional medium requires this separation to be zero. However, in practice, due to the physical dimensions of sources, this is not possible. Thus a compromise between source separation and the amount and physical extent of cancellation is sought.

Cancellation of noise in three-dimensional propagation is found to be dependent on geometry-related and source-related parameters. These include the difference in distance from the observer to the primary and secondary sources, the power ratio of the component waves, and the source frequency. A consideration of cancellation in terms of these parameters gives important insight into the conditions for cancellation as related to the geometry of the system and the characteristic nature of the secondary source. Furthermore, this leads to a three-dimensional description of the noise energy redistribution pattern in the form of zones of cancellation and reinforcement created through the process of field cancellation.

A description of the control structures employed in ANC systems reveals the extent of the practical and theoretical complexity of the required controller for cancellation of noise. A simple feedback control structure is found to require a controller of impractically large gain for optimum cancellation of the noise. Thus, a compromise between controller gain and the required level of cancellation is sought. Moreover, due to the feedback nature of this structure, care must be taken to ensure that stability conditions are satisfied over the frequency range of the noise.

The feedforward control structure requires a controller that is practically realizable for optimum cancellation of the noise. The dependence of the controller characteristics on system geometry reveals that certain combinations of the detector and observer locations relative to the primary and secondary sources lead to a controller with impractically large gain for optimum cancellation of the noise. However, it is found that these can be minimized by a proper positioning of the detector relative to the sources.

The acoustic feedback due to the secondary source radiation reaching the detector can cause the system to become unstable. Thus a stability analysis of the system is important at a design stage. It is found that system stability is closely related to system geometry. Therefore, the controller should be designed on the basis of the detector and observer locations corresponding to certain analytically defined regions of the medium which do not cause the system to become unstable.

The noise emitted by practical noise sources is generally of a broadband nature. In such cases the controller is required to produce a mirror image of each frequency component so that reduction over the broad frequency range of the noise is achieved. This means that the controller is required to have a continuous frequency-dependent transfer function. The dependence of the characteristics of the controller on measurable frequency-dependent parameters of the system (such as the characteristics of the source, transducers, and other electronic equipment, and response characteristics of the medium) makes it possible to obtain the required frequency-dependent characteristics of the controller and subsequently to implement a fixed digital controller. An ANC system employing such a controller is found to reduce the level of noise significantly over a broad frequency range.

The spectral content of noise of practical sources generally vary, for example, due to changes in operating conditions of the source, resulting in time-varying noise spectra. The effectiveness of a system employing a fixed digital controller can be significantly reduced under time-varying conditions. In such situations, the controller characteristics need to be updated in accordance with changes in the system so that a desired level of cancellation is achieved and maintained. This leads to the requirement of realizing the controller in an adaptive control framework; for example, using on-line controller identification to measure and calculate the required controller characteristics as a change in the system occurs. The recursive least squares identification method, for instance, can be used to give an estimate of the required controller transfer function. Development of the controller identification procedure within the ANC systems framework results in a self-tuning ANC system. Such a system is capable of maintaining the desired amount of cancellation over a broad frequency range under time-varying conditions as well as when the geometrical arrangement of the system is altered. The activation of self-tuning control can be achieved by a performance-dependent meta-level control (supervisory level). This self-tuning ANC algorithm as implemented on a digital signal processor performs well under time-varying conditions. Such an adaptive capability is essential and has great potential in the realization of successful ANC systems for real industrial applications.

The potential and usefulness of ANC systems in reducing unwanted acoustic noise emanating from a compact source in a three-dimensional non-dispersive propagation medium has been investigated and presented in this book, analytically as well as practically. The analytical procedure provides the basis of analysis for many other applications. For instance, a consideration of the problem of active noise attenuation in one-dimensional propagation (duct noise) can be considered in a similar manner with due consideration to the multi-mode problem of the duct. An extension of this to a more general case will provide the corresponding analytical consideration for a three-dimensional enclosure. In each case the problem will require a quantitative measure of cancellation in terms of the source-related and geometry-related parameters of the system, derivation of the required controller characteristics for a desired level of cancellation, and an analysis of the system from a stability point of view.

The problem of noise cancellation is closely related to the problem of vibration cancellation. Thus an extension to the development of an active vibration control system can readily be achieved. This will require modifications in the handling and interpretation of the transfer functions and signals involved.

The acoustic feedback in an ANC system due to the secondary signal reaching the detector can cause instability in the system and affects the complexity of the controller. This can be avoided by employing an indirect detection technique, that is, measuring a non-acoustic (coherent) signal. Many sources of acoustic noise, for instance, are found to vibrate at frequencies that are coherent with the noise. In such a case the noise can be detected through measuring the source vibration using an accelerometer. Flame noise is another example where the light is found to be coherent with the noise. Here the use of a piezo-electric film as a detector, for instance, can provide an indirect measure of the noise.

The transfer characteristics of detectors and actuators are further factors that contribute to the complexity of the required controller. The non-linearity and large attenuation in the characteristics of loudspeakers and microphones, especially in the low-frequency range, require the controller to have impractically large gain in this frequency range. Thus, to achieve a desired level of cancellation of the noise over an extended frequency range, improved microphones and loudspeakers with linear transfer characteristics are required.

A feature of many practical noise sources is that they are not compact. The physical dimensions of the source are appreciable proportions of the wavelength of the emitted noise, for example, in a high-power transformer, or the source may be thought of as composed of a number of sub-sources, such as in a commercial vehicle with the engine, exhaust, tyres, and so on, contributing to the overall noise level. A solution to such a case would be to treat the source as composed of a number of compact sources and use a multiple-detector and/or multiple-source configuration incorporating the methodology presented as the design basis. This will require a careful analysis of the problem considering the interaction between the sources resulting in multiple feedback loops that can cause the system to become unstable. Moreover, the separation of the sources playing a major role in redistributing the noise pattern in the medium requires an equal amount of attention. The design procedure on the basis of a fixed system geometry in a time-invariant environment in three dimensions results in a multi-input, multi-output controller composed of a number of sub-controllers each having frequency-dependent transfer characteristics dependent on frequency-dependent parameters within the system. The transfer characteristics of such a controller can be measured and will need to be realized in a multi-path structure, each path corresponding to a sub-controller and providing the required signal to a corresponding secondary source.

An extension of the above to include distributed sources with time-varying spectra by developing the concept of adaptive control within the ANC systems framework in a manner similar to the single-input single-output case will provide a solution to the general case of broadband, time-varying distributed sources of noise. This will require the simultaneous on-line identification of the transfer characteristics of the sub-controllers and the subsequent realization of these characteristics as a multi-input, multi-output controller and the addition of a supervisory level to activate the adaptation process.

It is evident that the demand for sophisticated and powerful digital signal processing devices varies in accordance with the system complexity. Important features of consideration in a digital signal processor for ANC applications include processing speed, dynamic range of computation, number representation, and extent of data and program memory space. Processing speed to comply with the real-time implementation requirements of digital controllers is important in ANC applications. Computational accuracy which is closely related to the processor's dynamic range of computation, number representation, and so on, is an equally significant feature required of a digital signal processor. Furthermore, as the system grows in complexity an increase in the program and data memory space will be required.

This book has presented and verified the possibility of designing and implementing a successful ANC system for real practical applications. To make this a practical reality developments and experimentation on a number of different industrial applications are required. Although this book is complete in its own right, it is felt that the end of the road has not yet been reached; much remains to be done, particularly in realistic industrial approaches before the 'sound of silence' can be heard.

APPENDIX A

The locus of the constant distance ratio

Theorem A.1 *Let P and S be two fixed points in a three-dimensional space with a distance d apart from each other and let T be an arbitrary point in this space. If the ratio of the distances PT and ST is constant then the locus of points T defines:*

(1) *a sphere with centre located along the line PS if the distance ratio is not unity.*

(2) *a plane perpendicularly bisecting the line PS if the distance ratio is unity.*

Proof Consider the three-dimensional UVW-space of Fig. A.1 in which $P(0, 0, 0)$ and $S(u_s, v_s, w_s)$ represent two fixed points and $T(u, v, w)$ an arbitrary point. The distances PS, PT, and ST are respectively denoted by d, r_g, and r_h. Let the distance ratio r_g/r_h be denoted by a positive real number a:

$$\frac{r_g}{r_h} = a. \tag{A.1}$$

It follows from Fig. A.1 that

$$d = [u_s^2 + v_s^2 + w_s^2]^{1/2} \tag{A.2}$$

and

$$\begin{aligned} r_g &= [u^2 + v^2 + w^2]^{1/2} \\ r_h &= [(u - u_s)^2 + (v - v_s)^2 + (w - w_s)^2]^{1/2}. \end{aligned} \tag{A.3}$$

Substituting for r_g and r_h from equations (A.3) into equation (A.1), simplifying, and using equation (A.2) yields

$$(1 - a^2)u^2 + 2a^2u_su + (1 - a^2)v^2 + 2a^2v_sv + (1 - a^2)w^2 + 2a^2w_sw = a^2d^2. \tag{A.4}$$

This gives the locus of points in the three-dimensional UVW-space that corresponds to a particular distance ratio a.

Non-unity distance ratio If the distance ratio a is different from unity then equation (A.4), after completing squares and simplifying, yields

$$\left[u + \frac{a^2u_s}{1 - a^2}\right]^2 + \left[v + \frac{a^2v_s}{1 - a^2}\right]^2 + \left[w + \frac{a^2w_s}{1 - a^2}\right]^2 = \left[\frac{ad}{1 - a^2}\right]^2; \quad a \neq 1. \tag{A.5}$$

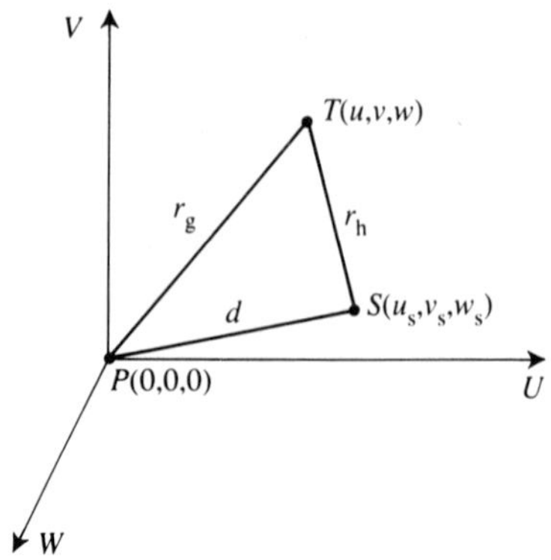

Fig. A.1 Fixed points P and S and arbitrary point T in three dimensions

This equation represents a sphere with radius $R = ad/|1 - a^2|$ and centre at $Q(-a^2u_s/(1 - a^2),\ -a^2v_s/(1 - a^2),\ -a^2w_s/(1 - a^2))$.

In order to prove that the centre of the sphere Q is located along the line PS, let the coordinates of this point be denoted by (u_q, v_q, w_q), unit vectors in the directions of the U-, V-, and W-axes, be denoted by **i**, **j**, and **k**, respectively, and unit vectors along the lines PS and PQ pointing from P towards points S and Q be denoted by $\boldsymbol{I}_{ps}$ and $\boldsymbol{I}_{pq}$, respectively. Thus,

$$u_q = -\frac{a^2u_s}{1 - a^2}$$

$$v_q = -\frac{a^2v_s}{1 - a^2} \qquad \text{(A.6)}$$

$$w_q = -\frac{a^2w_s}{1 - a^2}$$

and

$$\boldsymbol{I}_{ps} = \frac{u_s\mathbf{i} + v_s\mathbf{j} + w_s\mathbf{k}}{\sqrt{u_s^2 + v_s^2 + w_s^2}}$$

$$\boldsymbol{I}_{pq} = \frac{u_q\mathbf{i} + v_q\mathbf{j} + w_q\mathbf{k}}{\sqrt{u_q^2 + v_q^2 + w_q^2}} \qquad \text{(A.7)}$$

Substituting for u_q, v_q, and w_q from equations (A.6) into equation (A.7), simplifying, and using equation (A.2) yields

$$\boldsymbol{I}_{ps} = \frac{1}{d}(u_s\mathbf{i} + v_s\mathbf{j} + w_s\mathbf{k})$$

$$\boldsymbol{I}_{pq} = -\frac{|1 - a^2|}{1 - a^2}\frac{1}{d}(u_s\mathbf{i} + v_s\mathbf{j} + w_s\mathbf{k}) \qquad \text{(A.8)}$$

which implies that

$$\boldsymbol{I}_{pq} = \begin{cases} +\boldsymbol{I}_{ps} & \text{for } a > 1 \\ -\boldsymbol{I}_{ps} & \text{for } a < 1. \end{cases} \tag{A.9}$$

It follows from equations (A.9) that the centre of the sphere Q is located along the line PS and, specifically, if P is chosen as reference then for $a > 1$ the centre is located on the portion of PS corresponding to points away from P in the direction of $\boldsymbol{I}_{ps}$, whereas for $a < 1$ the centre of the sphere will be on the portion of PS corresponding to points away from P in the direction of $-\boldsymbol{I}_{ps}$. In either of these situations, it follows from equation (A.6) that the centre of the sphere lies outside the range (P, S). This is proved as follows.

Let the distance between points P and Q be denoted by r_{pq} and that between points Q and S be denoted by r_{sq}; thus

$$r_{pq} = \sqrt{u_q^2 + v_q^2 + w_q^2}$$

$$r_{sq} = \sqrt{(u_q - u_s)^2 + (v_q - v_s)^2 + (w_q - w_s)^2}.$$

Substituting for u_q, v_q, and w_q from equations (A.6) into the above, using equation (A.2), and simplifying yields

$$r_{pq} = \frac{a^2 d}{|1 - a^2|}$$

$$r_{sq} = \frac{d}{|1 - a^2|} \tag{A.10}$$

from which it follows that

$$r_{pq} > r_{sq} \quad \text{and} \quad r_{pq} > d \quad \text{for } a > 1$$

$$r_{pq} < r_{sq} \quad \text{and} \quad r_{sq} > d \quad \text{for } a < 1.$$

This proves that Q is always outside the range (P, S).

Let the line passing through points P and S intersect the sphere in equation (A.5) at point N (u_n, v_n, w_n) with distances r_{pn} and r_{sn}, respectively, relative to points P and S, and let the unit vectors pointing towards N from points P and S be $\boldsymbol{I}_{pn}$ and $\boldsymbol{I}_{sn}$, respectively; thus

$$r_{pn} = \sqrt{u_n^2 + v_n^2 + w_n^2}$$

$$r_{sn} = \sqrt{(u_n - u_s)^2 + (v_n - v_s)^2 + (w_n - w_s)^2} \tag{A.11}$$

and

$$\boldsymbol{I}_{pn} = \frac{u_n\mathbf{i} + v_n\mathbf{j} + w_n\mathbf{k}}{r_{pn}}$$

$$\boldsymbol{I}_{sn} = \frac{(u_n - u_s)\mathbf{i} + (v_n - v_s)\mathbf{j} + (w_n - w_s)\mathbf{k}}{r_{sn}}. \tag{A.12}$$

Since N is a point along the line PS the vectors $\boldsymbol{I}_{\text{pn}}$ and $\boldsymbol{I}_{\text{sn}}$ are either pointing in the same direction or in opposite directions. Thus it follows from equations (A.12) that

$$\frac{|u_{\text{n}}|}{r_{\text{pn}}} = \frac{|u_{\text{n}} - u_{\text{s}}|}{r_{\text{sn}}}$$
$$\frac{|v_{\text{n}}|}{r_{\text{pn}}} = \frac{|v_{\text{n}} - v_{\text{s}}|}{r_{\text{sn}}} \tag{A.13}$$
$$\frac{|w_{\text{n}}|}{r_{\text{pn}}} = \frac{|w_{\text{n}} - w_{\text{s}}|}{r_{\text{sn}}}.$$

Since $r_{\text{pn}}/r_{\text{sn}}$ represents the distance ratio a, defined in equation (A.1), equations (A.13) can be written as

$$u_{\text{n}}^2 = a^2(u_{\text{n}} - \text{u}_{\text{s}})^2$$
$$v_{\text{n}}^2 = a^2(v_{\text{n}} - v_{\text{s}})^2 \tag{A.14}$$
$$w_{\text{n}}^2 = a^2(w_{\text{n}} - w_{\text{s}})^2.$$

Solving the above for u_{n}, v_{n}, and w_{n} yields

$$u_{\text{n}} = \left(\frac{-a^2 \pm a}{1 - a^2}\right)u_{\text{s}}$$
$$v_{\text{n}} = \left(\frac{-a^2 \pm a}{1 - a^2}\right)v_{\text{s}} \tag{A.15}$$
$$w_{\text{n}} = \left(\frac{-a^2 \pm a}{1 - a^2}\right)w_{\text{s}}.$$

It follows from equations (A.15) that the line passing through points P and S and the sphere in equation (A.5) intersect at two points E and F with coordinates $(u_{\text{e}}, v_{\text{e}}, w_{\text{e}})$ and $(u_{\text{f}}, v_{\text{f}}, w_{\text{f}})$, respectively:

$$u_{\text{e}} = \frac{a}{1 + a}u_{\text{s}}$$
$$v_{\text{e}} = \frac{a}{1 + a}v_{\text{s}} \tag{A.16}$$
$$w_{\text{e}} = \frac{a}{1 + a}w_{\text{s}}$$

and

$$u_{\text{f}} = -\frac{a}{1 - a}u_{\text{s}}$$
$$v_{\text{f}} = -\frac{a}{1 - a}v_{\text{s}} \tag{A.17}$$
$$w_{\text{f}} = -\frac{a}{1 - a}w_{\text{s}}.$$

Equations (A.16) and (A.17) imply that E is a point that is always located inside the range (P, S) and F is outside this range. In particular, if $a > 1$ points E and F are nearer to point S, whereas if $a < 1$ then points E and F are nearer to point P. If the distances from point E to points P and S are denoted by r_{pe} and r_{se}, respectively and the distances from point F to points P and S are denoted by r_{pf} and r_{sf}, respectively, then it follows from equations (A.2), (A.16), and (A.17) that

$$r_{\mathrm{pe}} = \frac{ad}{1+a}$$
$$r_{\mathrm{se}} = \frac{d}{1+a} \tag{A.18}$$

and

$$r_{\mathrm{pf}} = \frac{ad}{|1-a|}$$
$$r_{\mathrm{sf}} = \frac{d}{|1-a|}. \tag{A.19}$$

Unity distance ratio If $a = 1$ then equation (A.1) yields

$$r_{\mathrm{g}} = r_{\mathrm{h}}. \tag{A.20}$$

Substituting for r_{g} and r_{h} from equations (A.3) into equation (A.20), simplifying, and using equation (A.2) yields

$$\frac{u}{\left(\dfrac{d^2}{2u_{\mathrm{s}}}\right)} + \frac{v}{\left(\dfrac{d^2}{2v_{\mathrm{s}}}\right)} + \frac{w}{\left(\dfrac{d^2}{2w_{\mathrm{s}}}\right)} = 1. \tag{A.21}$$

This represents a plane surface which intersects the U-, V-, and W-axes, respectively, at the points $(d^2/2u_{\mathrm{s}}, 0, 0)$, $(0, d^2/2v_{\mathrm{s}}, 0)$, and $(0, 0, d^2/2w_{\mathrm{s}})$.

The direction of the plane (A.21) is represented by a unit vector at right angle to the surface and pointing outward from the surface. Let such a unit vector be denoted by $\boldsymbol{I}_{\mathrm{s}}$. Simplifying equation (A.21) yields

$$2u_{\mathrm{s}}u + 2v_{\mathrm{s}}v + 2w_{\mathrm{s}}w - d^2 = 0.$$

Let the left-hand side of the above equation be denoted by some variable Z:

$$Z = 2u_{\mathrm{s}}u + 2v_{\mathrm{s}}v + 2w_{\mathrm{s}}w - d^2 \tag{A.22}$$

so that as Z varies from $-\infty$ to $+\infty$ equation (A.22) defines an infinite set of plane surfaces parallel to the plane in equation (A.21), thus the unit vector $\boldsymbol{I}_{\mathrm{s}}$

is given by

$$\boldsymbol{I}_{\mathrm{s}} = \frac{\dfrac{\partial Z}{\partial u}\mathbf{i} + \dfrac{\partial Z}{\partial v}\mathbf{j} + \dfrac{\partial Z}{\partial w}\mathbf{k}}{\sqrt{\left[\dfrac{\partial Z}{\partial u}\right]^2 + \left[\dfrac{\partial Z}{\partial v}\right]^2 + \left[\dfrac{\partial Z}{\partial w}\right]^2}} \tag{A.23}$$

where $\partial/\partial u$, $\partial/\partial v$, and $\partial/\partial w$ respectively denote the partial derivatives with respect to u, v, and w. Substituting for Z from equation (A.22) into equation (A.23), simplifying, and using equation (A.2) yields

$$\boldsymbol{I}_{\mathrm{s}} = \frac{u_{\mathrm{s}}}{d}\mathbf{i} + \frac{v_{\mathrm{s}}}{d}\mathbf{j} + \frac{w_{\mathrm{s}}}{d}\mathbf{k} \tag{A.24}$$

comparison of which with equations (A.8) gives

$$\boldsymbol{I}_{\mathrm{s}} = \boldsymbol{I}_{\mathrm{ps}}. \tag{A.25}$$

Equation (A.25) implies that the line PS is perpendicular to the plane surface in equation (A.21). Moreover, it follows from equation (A.20) that the point of intersection of the plane surface and the line PS is equidistant from points P and S. Therefore the plane in equation (A.21) perpendicularly bisects the line PS.

Corollary A.1 *Let P and S be two fixed points in a two-dimensional UV-space with a distance d apart from one another and let T be an arbitrary point in this space. If the ratio of the distances PT and ST is constant then the locus of points T defines*

(1) *a circle with centre located along the line PS if the distance ratio is not unity;*
(2) *a straight line perpendicularly bisecting the line PS if the distance ratio is unity.*

Proof The proof of this corollary follows directly from the proof of Theorem A.1 by eliminating the W-axis (equating the w-coordinates in equations (A.2)–(A.24) to zero). In doing so the corresponding results are obtained as follows.

Substituting for $w = 0$ and $w_{\mathrm{s}} = 0$ in equation (A.5) yields

$$\left[u + \frac{a^2 u_{\mathrm{s}}}{1 - a^2}\right]^2 + \left[v + \frac{a^2 v_{\mathrm{s}}}{1 - a^2}\right]^2 = \left[\frac{ad}{1 - a^2}\right]^2; \qquad a \neq 1. \tag{A.26}$$

This represents a circle with radius $R = ad/|1 - a^2|$ and centre at $Q(-a^2 u_{\mathrm{s}}/(1 - a^2),\ -a^2 v_{\mathrm{s}}/(1 - a^2))$ which is located along the line PS.

Substituting for $w = 0$ and $w_s = 0$ into equation (A.21) yields

$$\frac{u}{\left(\dfrac{d^2}{2u_s}\right)} + \frac{v}{\left(\dfrac{d^2}{2v_s}\right)} = 1; \qquad a = 1. \tag{A.27}$$

This represents a straight line intersecting the U- and V-axes respectively at points $(d^2/2u_s, 0)$ and $(0, d^2/2v_s)$.

From Fig. A.1 the line passing through points P and S is given by

$$v = \frac{v_s}{u_s} u. \tag{A.28}$$

Solving equations (A.27) and (A.28) simultaneously and using equation (A.2) with $w_s = 0$ gives the coordinates of the point of intersection of the two lines as $(u_s/2, v_s/2)$, the distance of which from point P is found to be equal to $d/2$, that is, midway between points P and S. The slopes m_1 and m_2, respectively, of the lines in equations (A.27) and (A.28) are

$$m_1 = -\frac{u_s}{v_s}$$

$$m_2 = +\frac{v_s}{u_s}$$

that is,

$$m_1 = -\frac{1}{m_2}$$

which implies that the two lines make a 90° angle at their point of intersection. Therefore, the line in equation (A.27) perpendicularly bisects the line PS.

APPENDIX B

The locus of the constant distance difference

Consider two fixed points P and S in a three-dimensional UVW-space with a distance d relative to one another and coordinates $(+d/2, 0, 0)$ and $(-d/2, 0, 0)$, respectively, and an arbitrary point $T(u, v, w)$ at a distance r_g relative to P and a distance r_h relative to S, as shown in Fig. B.1. The locus of points T for which the (distance) difference Δr between r_g and r_h is constant is investigated. It follows from Fig. B.1 that this difference can vary only between $-d$ and $+d$:

$$\Delta r = r_g - r_h = \zeta d \qquad \text{for } |\zeta| \leq 1. \tag{B.1}$$

From Fig. B.1 the distances r_g and r_h are

$$\begin{aligned} r_g &= \left[\left(u - \frac{d}{2}\right)^2 + v^2 + w^2\right]^{1/2} \\ r_h &= \left[\left(u + \frac{d}{2}\right)^2 + v^2 + w^2\right]^{1/2}. \end{aligned} \tag{B.2}$$

Substituting for r_g and r_h from equations (B.2) into equation (B.1) yields

$$\left[\left(u - \frac{d}{2}\right)^2 + v^2 + w^2\right]^{1/2} = \zeta d + \left[\left(u + \frac{d}{2}\right)^2 + v^2 + w^2\right]^{1/2}.$$

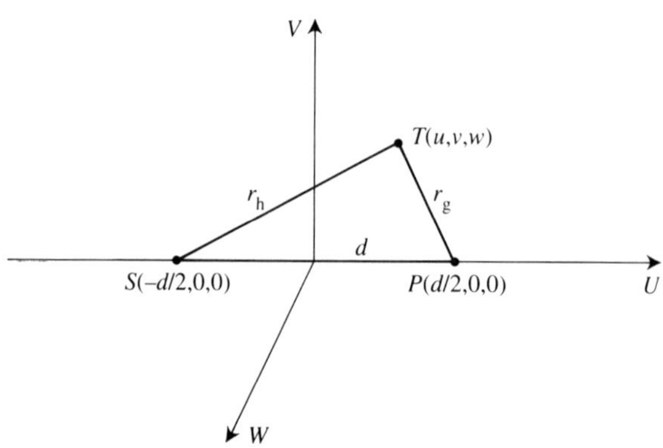

Fig. B.1 Fixed points P and S and arbitrary point T in three dimensions

Squaring both sides of this equation and rearranging yields

$$-\zeta^2 d^2 - 2du = 2\zeta d\left[\left(u + \frac{d}{2}\right)^2 + v^2 + w^2\right]^{1/2}.$$

Squaring both sides of the above equation and simplifying yields the locus of points T in Fig. B.1 for which the difference in its distance, Δr, from points P and S is constant as

$$4(1-\zeta^2)u^2 - 4\zeta^2 v^2 - 4\zeta^2 w^2 = d^2\zeta^2(1-\zeta^2). \tag{B.3}$$

In order to get a clear picture of the locus under consideration it is required to study equation (B.3) for the three cases of (*a*) $\zeta = 0$, (*b*) $|\zeta| < 1$ and $\zeta \neq 0$, and (*c*) $|\zeta| = 1$.

$\zeta = 0$ It follows from equation (B.1) that $\zeta = 0$ corresponds to the arbitrary point T in Fig. B.1 being equidistant from points P and S. Substituting zero for ζ into equation (B.3) and simplifying yields the locus of such points as

$$u = 0; \qquad \zeta = 0. \tag{B.4}$$

This defines a plane surface that coincides with the VW-plane in Fig. B.1, and perpendicularly bisects the line PS.

$1 > |\zeta| \neq 0$ It follows from equation (B.1) that this situation corresponds with the arbitrary point T in Fig. B.1 defining a locus for which the distance difference Δr is less than d in magnitude and different from zero. In this case equation (B.3), after simplification, yields

$$\frac{u^2}{\zeta^2\frac{d^2}{4}} - \frac{v^2}{(1-\zeta^2)\frac{d^2}{4}} - \frac{w^2}{(1-\zeta^2)\frac{d^2}{4}} = 1; \qquad 1 > |\zeta| \neq 0 \tag{B.5}$$

which, for ζ as a parameter within the ranges (0, 1) and $(-1, 0)$, defines a family of hyperbolic surfaces, in the three-dimensional UVW-space of Fig. B.1. For $0 < \zeta < 1$ the focus of the hyperbolic surface is at point S, whereas if $-1 < \zeta < 0$ the focus is at point P.

$|\zeta| = 1$ It follows from equation (B.1) that $|\zeta| = 1$ corresponds to the arbitrary point T in Fig. B.1 defining points along the line PS and outside the range (P, S). Substituting unity for ζ^2 into equation (B.3) and simplifying yields the locus of such points as

$$v^2 + w^2 = 0; \qquad |\zeta| = 1. \tag{B.6}$$

This defines two lines in the three-dimensional UVW-space of Fig. B.1 that extend, respectively, from points P and S outwards along the line PS to infinity.

Equations (B.4), (B.5), and (B.6) together completely define the locus of points T in Fig. B.1 for which the distance difference Δr is constant. It follows from these equations that the locus in equation (B.3), for ζ as a parameter, forms a family of hyperbolic surfaces with foci either at point P or point S. If $\zeta = -1$ the locus forms a line along PS extending outwards from point P. As ζ increases so that $-1 < \zeta < 0$ the locus forms a hyperbolic surface with focus at point P. For $\zeta = 0$ the locus becomes a plane surface that perpendicularly bisects the line PS. Increasing ζ so that $0 < \zeta \leq 1$ the locus forms an exact mirror image of the above about the plane for $\zeta = 0$.

APPENDIX C

The locus of the constant distance ratio–constant distance difference

Consider two fixed points P and S in a three-dimensional UVW-space with a distance d relative to one another and coordinates $(+d/2, 0, 0)$ and $(-d/2, 0, 0)$, respectively, and an arbitrary point $T(u, v, w)$ at a distance r_g relative to P and a distance r_h relative to S, as shown in Fig. C.1. The locus of constant distance ratio–constant distance difference is defined as the intersection of the locus of points T for which both the ratio r_g/r_h and the difference $r_g - r_h$ are constant.

Let the distance ratio r_g/r_h and distance difference $r_g - r_h$ be denoted by a and b, respectively:

$$\begin{aligned} \frac{r_g}{r_h} &= a, \qquad a \geq 0 \\ r_g - r_h &= b, \qquad |b| \leq d. \end{aligned} \tag{C.1}$$

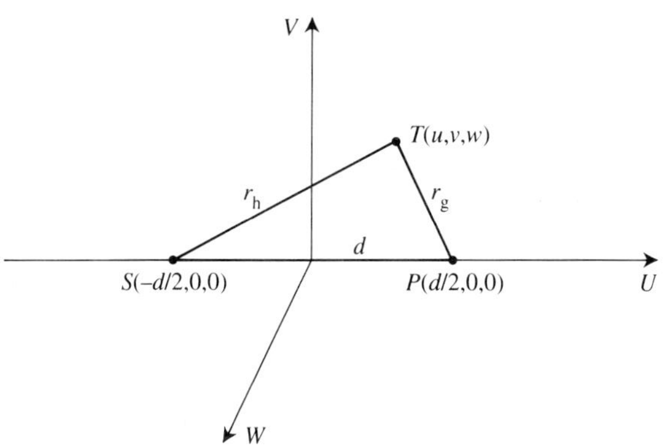

Fig. C.1 Fixed points P and S and arbitrary point T in three dimensions

From Fig. C.1 the distances r_g and r_h are

$$r_g = \left[\left(u - \frac{d}{2}\right)^2 + v^2 + w^2\right]^{1/2}$$
$$r_h = \left[\left(u + \frac{d}{2}\right)^2 + v^2 + w^2\right]^{1/2} \tag{C.2}$$

Substituting for r_g and r_h from equations (C.2) into the first of equations (C.1) and simplifying yields

$$(1 - a^2)\left(u^2 + v^2 + w^2 + \frac{d^2}{4}\right) - (1 + a^2)du = 0. \tag{C.3}$$

Equation (C.3) defines a plane surface in the three-dimensional UVW-space of Fig. C.1 that perpendicularly bisects the line PS if $a = 1$, whereas if $a \neq 1$ it defines a sphere with centre along the line PS:

$$u = 0 \qquad \text{for } a = 1$$
$$\left(u - \frac{1 + a^2}{1 - a^2}\frac{d}{2}\right)^2 + v^2 + w^2 = \frac{ad^2}{(1 - a^2)^2} \qquad \text{for } 0 \leq a \neq 1. \tag{C.4}$$

Substituting for r_g and r_h from equations (C.2) into the second of equations (C.1) and simplifying yields

$$4(d^2 - b^2)u^2 - 4b^2v^2 - 4b^2w^2 - b^2(d^2 - b^2) = 0. \tag{C.5}$$

Equation (C.5) defines a plane surface in the three-dimensional UVW-space of Fig. C.1 that perpendicularly bisects the line PS if $b = 0$, whereas if $b \neq 0$ it defines a hyperbolic surface:

$$u = 0 \qquad \text{for } b = 0$$
$$\frac{u^2}{\frac{b^2}{4}} - \frac{v^2}{\frac{d^2 - b^2}{4}} - \frac{w^2}{\frac{d^2 - b^2}{4}} = 1 \qquad \text{for } d \geq |b| \neq 0. \tag{C.6}$$

It follows from the above that the locus of constant distance ratio–constant distance difference is defined as the intersection of the loci given by equations (C.4) with those given by equations (C.6). These imply that if $b = 0$ and $a = 1$ then the locus is a plane surface in the three-dimensional UVW-space of Fig. C.1 that perpendicularly bisects the line PS:

$$u = 0; \qquad a = 1 \qquad \text{and} \qquad b = 0 \tag{C.7}$$

whereas if $a \neq 1$ and $b \neq 0$ the locus is given by the intersection of the spherical surface in equations (C.4) with the hyperbolic surface in equations (C.6). Such an intersection will result in a circle on a plane surface that is perpendicular to a line along PS (here the U-axis). The centre of the circle (hereafter referred to as the constant a–constant b circle) is also along this line.

If the u-coordinate of the constant a–constant b circle is denoted by u_c then, for the intersection of the sphere in equations (C.4) and the hyperbolic surface in equations (C.6) to exist, it is necessary that u_c be outside the range given by the intersection of the hyperbolic surface with the U-axis. The u-coordinate of the point at which the hyperbolic surface crosses the U-axis is obtained, after substituting zero for v and w into equation (C.6), as $u = b/2$. Therefore, the necessary condition for the constant a–constant b circle to exist for a given distance ratio a and distance difference b is

$$\begin{aligned} u_c &\leq \frac{b}{2} \qquad \text{for } -d \leq b < 0 \\ u_c &\geq \frac{b}{2} \qquad \text{for } 0 < b \leq +d. \end{aligned} \tag{C.8}$$

Substituting for $v^2 + w^2$ from the second of equations (C.6) into the second of equations (C.4) and simplifying yields the following relation for the coordinate $u = u_c$ of their intersection

$$u_c^2 - \frac{b^2}{d}\frac{1+a^2}{1-a^2}u_c + \frac{b^4}{4d^2} = 0 \tag{C.9}$$

solution of which for u_c satisfying the necessary condition in equation (C.8) yields

$$u_c = \frac{b^2}{2d}\frac{1+a}{1-a}. \tag{C.10}$$

Substituting for $u = u_c$ from equation (C.10) into the second of equations (C.6) and simplifying yields

$$v^2 + w^2 = \frac{d^2}{4}\left[1 - \frac{b^2}{d^2}\right]\left[\frac{b^2}{d^2}\left(\frac{1+a}{1-a}\right)^2 - 1\right]; \qquad a \neq 1 \text{ and } b \neq 0. \tag{C.11}$$

This represents a circle on a plane surface that is parallel to the VW-plane, with centre along the U-axis at a location given by equation (C.10).

Equations (C.7) and (C.11) define the locus of points T in the three-dimensional UVW-space of Fig. C.1 for which the distance ratio r_g/r_h as well as the distance difference $r_g - r_h$ is constant. The former is a plane surface that perpendicularly bisects the line PS and the latter is a circle, with centre along the line PS, on a plane surface that is parallel to the VW-plane.

Index